Naturalists' Handbooks 24

Ants

2nd edition

GARY J. SKINNER & ANDREW P. JARMAN

Pelagic Publishing

Second edition published in 2025 by
Pelagic Publishing
20–22 Wenlock Road
London N1 7GU, UK

www.pelagicpublishing.com

First edition published in 1996 by
The Richmond Publishing Co. Ltd

Naturalists' Handbooks
Series editor: William D.J. Kirk

A CIP catalogue record for this book is
available from the British Library

ISBN 978-1-78427-304-0 Pbk
ISBN 978-1-78427-305-7 ePub
ISBN 978-1-78427-306-4 PDF

https://doi.org/10.53061/MIHN9342

EU Authorised Representative:
Easy Access System Europe – Mustamäe tee 50,
10621 Tallinn, Estonia, gpsr.requests@easproject.com

Cover images, clockwise from top: *Formica rufa* workers,
Lasius flavus workers with the myrmecophilous
woodlouse *Platyarthrus hoffmannseggii*,
Tetramorium caespitum workers with brood.

Typeset by BBR Design, UK

5 4 3 2

Printed in the Czech Republic by Finidr.

Contents

Editor's preface

Ants occur almost everywhere. We usually just ignore them or sometimes try to get rid of them. However, ants are fascinating because of the way they function as colonies and manage cooperation and conflict among many workers. Ants are also important in the environment because they are ecosystem engineers and generalist predators – for example, they modify the soil and affect the abundance of many other organisms.

This *Naturalists' Handbook* gives an introduction to the natural history, biology, ecology and conservation of ants and provides advice on how to study them, together with identification keys to the species in Britain, Ireland and the Channel Islands.

Ants are highly suitable for investigation because of their ubiquity and abundance. They can be found in gardens, parks, grassland, woods and even under paving stones or infesting buildings. Colonies can be kept indoors and observed easily throughout the year in ant farms or formicaria. There is still much to be discovered about their biology and ecology. Records of which species occur where can be submitted to national recording schemes. Such contributions are useful because they improve our knowledge of species distributions and this can then be used to show the effects of climate change, insect decline or changes in land-use.

The first edition of *Ants* was published 28 years ago, in 1996. This second edition is considerably expanded with many more colour illustrations and revised identification keys. It offers an updated and more detailed coverage of the subject. *Ants* complements several other titles in this series: *Animals under logs and stones* (No. 22), *Studying invertebrates* (No. 28) and *Aphid predators* (No. 11).

William D.J. Kirk
December 2024

About Naturalists' Handbooks

Naturalists' Handbooks encourage and enable those interested in natural history to undertake field study, make accurate identifications and produce original contributions to research. A typical reader may be studying natural history at sixth-form or undergraduate level, carrying out species/habitat surveys as an ecological consultant, undertaking academic research or simply developing a deeper understanding of natural history.

About the authors

Gary J. Skinner began studying chemistry but soon decided to swap to biology and obtained a degree in 1972. He then went on to study the wood ant *Formica rufa* for his PhD. After this he had a career in teaching, until his retirement in 2008. In the 1980s whilst on a trip to Skomer he saw a little book on British buttercups and thought 'I could do one on ants'. That was published in 1987 and its success led him to think about a Naturalists' Handbook, the first edition of which came out in 1996. Teaching in a boarding school was very demanding but he managed to fit in some ant observing, especially in the 1990s when he undertook survey work in the north-west of England during a sabbatical term. He has written extensively across the biological sciences and was editor of the magazine *Catalyst* for 10 years until 2017. In retirement he has continued to write and mark GCSE and A level examinations.

Andrew P. Jarman has had a lifelong fascination with ants. He claims that his earliest memories are of discovering that there were three types of ants in his parents' garden (black, red and yellow) before he even learnt to walk. Over the decades since, he has accumulated an extensive field and taxonomic knowledge of ants in Britain, as well as gaining a working knowledge of the ant faunas of continental Europe, Central and North America, and South-East Asia. He has been a member of the Bees, Wasps and Ants Recording Society since its earliest days and was a past committee member. In his day job, he is a lecturer and researcher in biomedical sciences, specialising in the neurobiology of the laboratory fruit fly, *Drosophila melanogaster*.

Acknowledgements

Even a relatively small book such as this has relied on a huge number of people to help us put it together and we are very grateful to them all. We hope that will cover any whom we forget to thank in the following!

One of the challenges for GS during lockdowns was to get photographs of ants doing things, as opposed to set specimens. It only worked at all due to the generosity of many people, both with their time and their expertise. GS would like to thank, roughly in the order of the ants they helped him with, the following. Ross Johnston and his son, Xander 'antboy' Johnston, neither of whom he managed to meet but who guided him via email unerringly to nests of *Formica sanguinea* and *Formica aquilonia*, the latter in the hope of seeing both it and the elusive SGA (shining guest ant or *Formicoxenus nitidulus*). Murdo Macdonald helped him, again from afar, to locate good sites for *Formica exsecta* and *F. lugubris* on the same Aviemore trip. Next, Kevin Henry gave freely of his time in Carmarthenshire where we sought out *Formica picea*, with great success. Once on Guernsey, towards the end of his travels, he was ably assisted by Andy Marquis in locating *Formica pratensis* on the southern cliffs, together with the amazing *Strumigenys perplexa*, the latter meticulously sieved by the indomitable Andy from some road sweepings. Continuing the theme of help from afar, Ian Beavis pulled out all the stops via emails to direct him to 'angry corner' on St Martin's, Isles of Scilly, where he found any number of *Formica rufibarbis*. Scotty Dodd and Stephen Carroll helped with updates on *F. rufibarbis* on the mainland. He finally found SGA under the expert guidance of Elva Robinson, although this was again from afar as, when he went into the field with her, none were seen. Elva also helped him find *Lasius neglectus* at Kirk Smeaton. GS remembered seeing *Lasius fuliginosus* years ago on the Formby dunes, but did not want to go there without some assurance that it is still there, this species having a reputation of moving along. Sure enough, Chris Hunter went out to look on the strength of a mention in a 2009 report and, long story short, he followed and found this lovely ant. *Myrmecina graminicola* proved elusive to him but Mike Fox gets it in his garden when he puts out an orange, and so he got to see some when he met up with Mike to look at ants in Gunnersbury Triangle. Sadly, we did not find *Myrmica schencki* turrets there. Again from a distance, Alistair Kirk was very helpful and led him to be able to find *Stenamma debile*, again a first for him in the wild. Craig Edwards at Dungeness kindly organised permission to search for the two *Ponera* species there, but he did not locate either. However, he did get some great images of *Tetramorium caespitum* and *Temnothorax albipennis*. Special thanks go to Tony Hunter and Gary Hedges at the World Museum in Liverpool for welcoming GS and allowing him to use their focus-stacking setup to get most of the pictures in Chapter 4. Some of the material used came from the museum's own collection and that of Cedric Collingwood, which is housed in the museum. Many thanks also to Dmitri Luganov at Manchester Museum and Louis Lofthouse at Oxford University Museum of Natural History for allowing him to borrow and photograph some specimens that were not available in Liverpool. The remainder of the pictures of set specimens in Chapter 4 were supplied by AJ. Philip Ward advised on the current situation with regard to phylogeny, but it will probably change again by the time the book is a year old!

It has been a long journey with the intervention of COVID and we must thank William Kirk, our editor. GS extends especial thanks to William for sticking with him all this time and spotting all the things he did not and thus saving him a lot of potential embarrassment!

The keys here have been prepared by AJ especially for this book and this greatly benefited from the head start provided by Geoff Allen, who generously allowed reuse of figures from the first edition of this book. Prior to this, AJ had spent many years (on and off) drafting and developing identification keys, and over the years has relied on generous input of time and expertise by many people in this evolving process. Initially, Mark Shaw (formerly National Museums Scotland) had doggedly encouraged AJ to have a go at producing new keys to replace those in the old Royal Entomological Society Handbook on ants, and Barry Bolton (formerly Natural History Museum, London) generously made available his unpublished personal key to workers, which helped thinking about new approaches to some species. Bernhard Seifert, Phil Attewell and Murdo Macdonald imparted their extensive expertise of certain species. For this book specifically, AJ is grateful to Mike Fox for organising testing of the new keys by non-specialists, which resulted in some important beginner-friendly improvements. One of the testers, Dave Barnett is owed thanks for his insightful comments on an earlier draft of the worker key. Matt Hamer gave advice on the thorny issue of which introduced species to include.

For the images in the keys, AJ thanks Ashleigh Whiffin and Vladimir Blagoderov (National Museums Scotland) for access to their photomicrography equipment. All help and advice has been gratefully received, but not necessarily acted on appropriately, so any errors or deficiencies are of course the responsibility of AJ.

Finally, we wish to thank our spouses, Ann and Siân, for tolerating our obsessions, and even joining with us in them, searching for ants and nests all over these islands.

GS would also like to thank the following who have given permission for their images to be used. Their names appear with the images they supplied, in many cases free of charge.

- Fig. 1.8 Frank Carpenter
- Fig. 1.9 Nick Owens
- Fig. 2.1 & Fig. 4.3 Reprinted by permission of HarperCollins Publishers Ltd © 1977, M.V. Brian
- Fig. 2.7 & Fig. 6.16 Peter Furze
- Fig. 2.8 & Fig. 6.17 Grzegorz Wagner
- Fig. 2.22 Alex Wild
- Fig. 3.3 Jake Alagoa
- Fig. 3.12 Henry Disney
- Fig. 3.13 Jeremy Thomas
- Fig. 3.16 Philipp Hoenle
- Fig. 3.18 R. Dransfield & R. Brightwell, influentialpoints.com
- Fig. 4.1 Alexander Radchencko
- Fig. 6.13 Rosemary Winnall
- Fig. 7.7 Andy Marquis
- Fig. 7.9 NHBS Ltd
- Fig. 7.11 Mike Fox
- Fig. 7.17 Antstore, Germany

GS, AJ

1 Introduction

Most of us know what an ant looks like. We can confidently recognise a particular insect as an ant, even after a casual glance. How? What are the features that make an insect an ant? In early school biology lessons, we were taught that adult insects have wings. Ants are an exception to this almost universal rule. In our rapid identification of an ant, we notice the lack of wings. All ants are social and have a variety of different forms in their societies. Although only the workers are wingless, these are far more numerous and active than the winged reproductives and thus more likely to be seen by the casual observer.

Several other insects share this lack of wings, but ants have other features that help to set them apart. All ants belong to one family (the Formicidae) within an order of insects called the Hymenoptera. This order is divided into two major groups, the Apocrita and the Symphyta. The fundamental distinguishing feature is that the Apocrita (ants, bees, wasps and parasitic wasps) have a waist as an adult, whereas the Symphyta (sawflies) do not. Ants are Apocrita and thus are waisted. To distinguish ants from the few other insects that combine these two striking features – winglessness and a waist – look for the other two ant characteristics: one or two scale-like or bulbous swellings or nodes on the waist; and elbowed antennae. Any insect with no wings, a waist with nodes or scales and elbowed antennae is an ant (Fig. 1.1).

Do we really see all this in an animal less than one centimetre long when we recognise an ant at a picnic spot? No – the feature that gives the game away is the sociality, reflected in the sheer numbers. The fascination of ants lies in this sociality, and it is the organisation of ant societies, so similar to our own and yet so different, that will be one of the main areas of focus for this book.

Myrmecologists, people who study ants, agree that all ants belong to just the one family mentioned above, the Formicidae. This is divided into subfamilies. The number of subfamilies is currently thought to be 17, although there is variation amongst authorities. Table 1.1 shows the classification by Keller & Peeters (2020). In Britain and Ireland, the ant fauna is relatively sparse, with only four subfamilies out of a total world count of 17. In Ireland, two of the four (the Ponerinae and the Dolichoderinae) are each represented by only a single introduced species. Looking at species numbers, the situation appears even more disheartening for the aspiring

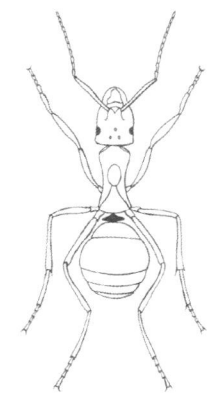

Fig. 1.1 A typical formicine ant, showing the one-segmented waist (after Skinner 1987)

Table 1.1 A world list of ant subfamilies (* British subfamilies, † Irish subfamilies) (based on Keller & Peeters 2020)

Subfamily	Number of genera/species	Notes
Agroecomyrmecinae	2/2	Only two genera, one in Central and South America and the other in West Africa. Only a few specimens seen, and little is known about their biology.
Amblyoponinae	9/144	Found throughout the tropics, with a few genera in North America, Europe and Asia. There are 31 species in Madagascar. In their biology they are army-ant like.
Aneuretinae	1/1	One species found only in Sri Lanka.
Apomyrminae	1/1	Found in West Africa, thought to feed on centipedes.
Dolichoderinae* †	28/842	Found throughout the world, but with most in the tropics. One of the big four subfamilies. Some very successful invasive species are in this subfamily, the best known being the Argentine ant. There are three species in Britain, one being the Argentine ant, which is introduced. There are suggestions that Argentine ants may have entered Ireland too.
Dorylinae	27/869	The best-known members of this subfamily are the army and driver ants of South America and Africa respectively.
Ectatomminae	12/304	Throughout the tropics (but not Africa).
Formicinae* †	52/4071	One of the big four subfamilies (25 species in Britain and 11 in Ireland). Wood ants, carpenter ants, honeypot ants and many familiar kinds are found in this subfamily.
Heteroponerinae	3/32	Main genus *Heteroponera* with 28 of the 32 species. Found in the neotropics and Australasia.
Leptanillinae	7/71	Minute subterranean species found in southern Europe across to Japan, with one species in Australia.
Martialinae	1/1	Only one genus, with one species known from just three specimens found in Brazil.
Myrmeciinae	2/94	A once widespread subfamily now restricted to Australia and New Zealand. The so-called bulldog or inch ants.
Myrmicinae* †	148/7802	One of the largest subfamilies and one of the big four. This is the subfamily represented by the greatest number of species in Britain, where there are 27. In Ireland there are eight.
Paraponerinae	1/1	One genus with just a single species which occurs in Central and South America: the bullet ant, so called because of the extreme pain caused by its sting.
Ponerinae* †	49/1399	A large subfamily of primitive ants found mainly in the tropics around the world. One of the big four, but with just five species in Britain and none so far found in Ireland.
Proceratiinae	3/166	Tiny, cryptic species seldom seen. Found throughout the tropics and into southern Europe.
Pseudomyrmecinae	3/243	The best known are those in the genus *Pseudomyrmica*, many of which inhabit specialist structures on trees, including the so-called acacia ants.

This book covers all the ant species found in Britain, Ireland and the Channel Islands. The Channel Islands have six or seven species that are not known to occur elsewhere in the region. The term Britain will be used throughout and qualified with any additional information about Ireland and the Channel Islands where relevant.

myrmecologist. Out of an estimated world list of nearly 16,000 species, the region covered by this book has about 60. However, this makes identification easier with only 60 to think about. Ants are fundamentally sun lovers, and our cold climate does not suit them. Nevertheless, a number of species survive well in our temperate latitudes and provide fascinating material for study. Some show behaviour almost as dramatic as that of the well-known tropical army ants and driver ants, so beloved of horror-story writers.

Two of the four subfamilies found in Britain are particularly poor in species. The Ponerinae are represented in Britain by only five species: *Hypoponera ergatandria*, *Hypoponera punctatissima*, *Hypoponera eduardi*, *Ponera coarctata* and *Ponera testacea*. None of these are very common. *Ponera coarctata* is truly native and is encountered sporadically in the south of England. *Ponera testacea* was raised to species status in 2003 and so far has been found in only about 20 sites in, again, the south of England, most notably at Dungeness (Figs 1.2, 1.3). The status of *H. punctatissima* is the subject of debate, but it does now seem to be widely accepted that this species is not confined to greenhouses and hothouses. A recent (2016) record was of workers carrying eggs found in a heap of horse manure near Cardiff in South Wales. *Hypoponera ergatandria*, on the other hand, is never seen away from buildings. *Hypoponera eduardi* is a very recent addition to the fauna, having turned up on the Isle of Wight and in London Zoo. The features of the Ponerinae subfamily and the five British species are given in Chapter 6.

The second small British subfamily, the Dolichoderinae, is represented by only two species: *Tapinoma erraticum* and *Tapinoma subboreale* (formerly, and briefly, known as *Tapinoma ambiguum*). *Tapinoma erraticum* is confined to south central England and is mainly coastal, but nests are easily overlooked,

Fig. 1.2 *Ponera coarctata* workers with larvae

Fig. 1.3 *Ponera testacea* worker

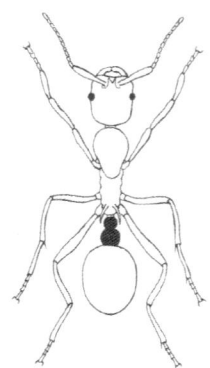

Fig. 1.4 A typical myrmicine ant, showing the two-segmented waist (after Skinner 1987)

wood ants

A group of closely related species within a subgenus of the genus *Formica*, which usually live in woodland and build thatched mound nests. They are also known as red wood ants, thatch ants or mound ants. There are three species in Britain (*F. rufa*, *F. aquilonia* and *F. lugubris*), one in the Channel Islands (*F. pratensis*) and two in Ireland (*F. lugubris* and *F. aquilonia*), of which *F. lugubris* is known to be genetically distinct from its counterpart in Britain (Breen 2014). Note that some other *Formica* species also build thatched mound nests and that 'wood ants' has been applied more loosely to include other *Formica* species in some older literature.

and it would be worth searching for them in likely habitats, mainly sandy heaths. *Tapinoma subboreale* has only recently been separated from *T. erraticum* (Seifert 2012) and records of *T. erraticum* may prove to be *T. subboreale*. The recency of the recognition that *T. erraticum* consists, in fact, of two species means that there is a rich area for investigation of the biological and ecological differences between the two. Neither of these species is found in Ireland.

Two other dolichoderines are sometimes encountered in Britain: the Argentine ant *Linepithema humile* (formerly *Iridomyrmex humile*) and the ghost ant *Tapinoma melanocephalum*. Both are so-called tramp species, invasive ants that are found widely across the world. The former is almost always encountered in heated buildings, although there is a large colony known from an outdoor site in Fulham, West London. Ghost ants are similarly associated with buildings.

Most of the ants found in both Britain and Ireland are members of the Myrmicinae and the Formicinae. These are easily distinguished from one another with a hand lens, or even sometimes without once you have 'got your eye in'. As mentioned, the waist of an ant bears scales or nodes. Formicine ants have just one (Fig. 1.1) whereas myrmicines have two (Fig. 1.4). It is useful to be able to distinguish these major groups in the field. Chapter 6 deals with identification in more detail.

1.1 Where to look for ants

As discussed, ants like warmth, and thus tend to be most abundant in warmer parts of Britain and Ireland (the south and west), or in warmer habitats (such as those on sandy soils). In warm weather during the active season, some species are easy to find, particularly when they are foraging for food. Thus, in woodland, it is always worth looking at the trunks of trees, on which workers may often be seen streaming up and down on their way to and from food sources amongst the foliage. In habitats dominated by shrubs and herbaceous plants, a careful search will often reveal ants looking for food or tending aphids.

Nearly all ants build nests, and these may be more obvious than the ants themselves. The hill ant, *Lasius flavus*, is a largely subterranean species and its vegetation-covered mound nests, which are 30 cm or more high, are much more likely to be spotted than the tiny, secretive workers (Fig. 1.5). Nests of the wood ants are very large domes made of pine needles and other materials, often more than one metre

Fig. 1.5 A mound nest of *Lasius flavus*. The camera lens on the left is over 30 cm long and shows the sheer scale of these nests.

high, and difficult to miss in the woodlands in which they occur (Fig. 1.6).

Some species are rarely found. In some cases, this is probably because they are truly rare. In others, it is almost certainly due to want of looking; this is an area in which useful contributions could be made by amateurs. Very little ant recording has been done in some parts of the country, especially in the north of England, much of southern Scotland, much of Ireland and parts of Wales (Fig. 1.7).

Some species are parasitic on others and do not form populous colonies. The distribution of these species (*Myrmica karavajevi*, *Myrmica hirsuta*, *Tetramorium atratulum* and *Strongylognathus testaceus*) is almost certainly imperfectly known. *Myrmica karavajevi* occurs only in nests of *Myrmica scabrinodis* and *Myrmica sabuleti*, and *Tetramorium atratulum* and *Strongylognathus testaceus* occur only in *Tetramorium*

Fig. 1.6 A thatched mound of *Formica aquilonia*

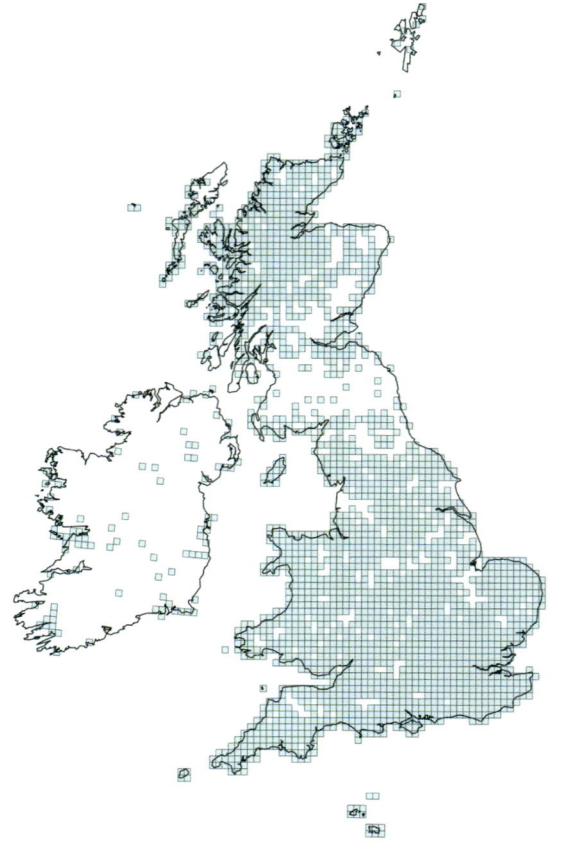

Fig. 1.7 The distribution of ant records received by BWARS (the Bees, Wasps and Ants Recording Society)

caespitum nests. Careful searching for these in host nests could be well worthwhile.

Some ants, although not parasitic in the same sense, are commonly associated with other species. These include *Formicoxenus nitidulus*, which lives within the nest mounds of *Formica rufa*, *Formica aquilonia* and *Formica lugubris*. It is considered difficult to find and is best looked for on dull, warm days. *Solenopsis fugax* is often associated with nests of *Lasius* and *Formica* species, and it is thought to prey upon their brood (the young stages). Some ants are very active and conspicuous, but several of our species are small and unobtrusive and need to be looked for very carefully. Such searching, coupled with accurate identification, could yield useful information about the distribution and biology of these relatively little-known species. There is an active scheme for recording the distribution of ants in Britain and Ireland (p. 303).

In many ways, ants are easy animals to study. The main attribute needed is patience. Long-term observations can be made by revisiting nests repeatedly over a period of days, months or even years. Such long-term records are relatively rare as not many have had the persistence to continue with such work. A few hours spent by a nest at intervals throughout the season can yield valuable information about food intake, activity patterns and energy input. Daily and seasonal variations in these can be related to weather and other factors. These studies, coupled with observations under more controlled conditions in formicaria (p. 289), can give insights into factors controlling behaviour.

Because of their abundance, ants affect the environment in which they live in various ways. The environment within the nest can be highly controlled and ants, like humans, are to some extent independent of prevailing external conditions. The availability of sophisticated, but quite inexpensive, data-logging devices makes it possible to monitor variations in conditions in ants' nests throughout the day and the season, but valuable information can also be gained with simple equipment such as a thermometer.

Further afield, ants may affect populations of their, often herbivorous, prey and therefore influence the growth of plants near nests. These effects can be revealed by comparing areas foraged by ants with ant-free areas. Prey populations can be monitored throughout the season, comparing situations where ants are present with others from which they are naturally absent, or have been excluded experimentally. Effects on prey may also be studied in laboratory colonies, where conditions can be controlled.

1.2 The evolutionary history of ants

The ants we see today have a long history. In 1967 a fossil ant called *Sphecomyrma freyi* was found in New Jersey, USA, preserved in amber of Cretaceous age, possibly 80 million years old (Fig. 1.8). The genus name means 'wasp ant' and the specific name is in honour of the couple who found the specimens, Mr and Mrs Frey. These early ants were probably not common; only about 1% of the insects found in Cretaceous amber are ants. It was 65 million years ago that the ants began to move towards the dominant role that they have now. Features of the anatomy of *Sphecomyrma freyi* link it with modern ants on the one hand and solitary wasps on the other.

There is controversy about the origin of primitive ants, but most researchers regard them as close to the tiphiid wasps. A modern tiphiid genus, thought to resemble ant ancestors, is *Methocha* (Fig. 1.9). Members of this genus prey on the larvae of tiger beetles (Cicindelidae). These larvae occupy a vertical burrow, from where they catch prey that walks past. *Methocha* females sting the larva, which is much bigger than them, and then lay an egg on it. The egg hatches after about four days and the wasp larva then feeds on the paralysed, but still living, tiger beetle larva. Wilson & Farish (1973), based on their observations of the species *Methocha stygia*, suggested that methochine wasps might well be close to the ancestors of ants. They noted that the typical behaviour described above could be varied. Some females transported prey and one wasp was seen to use a depression as a nest site. They concluded that 'this flexibility in response might

Fig. 1.8 One of the earliest ant fossils, *Sphecomyrma freyi*, preserved in amber (Frank Carpenter)

Fig. 1.9 A female *Methocha articulata*, the only species in this genus found in Britain; it is absent from Ireland (Nick Owens)

be a carry-over of the behaviour that surely existed in the ancestors of the methochine wasps, insects which were close to, if not identical with, the ancestors of the ants'.

A phylogeny (family tree) for ants is given in Fig. 1.10.

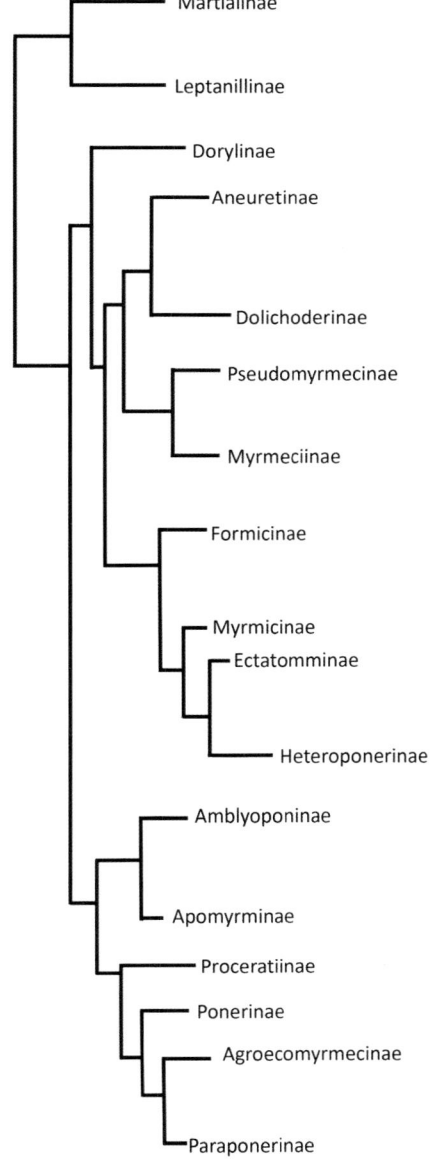

Fig. 1.10 A hypothetical phylogeny (family tree) for ants (after Keller & Peeters 2020)

1.3 Eusociality

There are solitary wasps and there are solitary bees, but there is no such thing as a solitary ant. All ant species are social. Their social behaviour qualifies as the most advanced type, known as eusociality, because they exhibit three important features: first there is a sterile worker caste, second there is overlap of generations and third the females cooperate with their sisters in caring for the eggs, larvae and pupae, collectively called the brood. This type of true social organisation is found in the ants, a few species of wasps and bees, all termites, two thrips species (the Australian *Kladothrips tepperi* and *Kladothrips habrus*) and a beetle (the Australian ambrosia beetle, *Austroplatypus incompertus*) among the insects. Eusociality is also seen in some snapping shrimps and the naked mole rat.

The evolutionary origin of eusociality presents biologists with a difficult problem. How can evolution favour a sterile worker caste if the workers never leave progeny of their own to carry their own genes into future generations? How can this apparently selfless behaviour arise? Darwin recognised this as a major problem with his own ideas; to him the presence of sterile workers represented 'one special difficulty, which at first appeared to me insupportable, and actually fatal to my whole theory' (Darwin 1859). He suggested that such sterility could arise because these forms are useful to the community, and that selection might act at the family level. This, however, is only a partial solution and does not begin to explain how sterility could arise in the first place. In modern terms, the problem can be put as follows, by Sudd & Franks (1987): 'Why should a female choose to stay in the parental nest rather than attempt to start her own family? How could genes that code for helping the reproduction of relatives become more abundant in future gene pools, if such genes also led to an individual producing fewer of its own direct descendants?' Davies *et al.* (2012) examine a number of hypotheses and a full account can be found in their excellent book. The following account follows Davies *et al.* closely.

Haplodiploidy

Males of ants (and other Hymenoptera) produce gametes (sperm) by mitosis rather than by meiosis, so all the sperm are genetically the same. This means that any daughter of that male will share all his genes with her sisters. There is a 50% chance that sisters will share any particular one of their mother's genes. Any female will have a relatedness

caste
A form of a species characterised by a particular role in the division of labour, such as queens and workers.

haplodiploidy
A sex determination system in which males are haploid (have one set of chromosomes) and develop from unfertilised eggs, and females are diploid (have two sets of chromosomes) and develop from fertilised eggs.

mitosis
A type of cell division in which the chromosomes are duplicated and then segregated into two nuclei. The cell then usually divides to give two identical daughter cells.

meiosis
A type of cell division in organisms that reproduce sexually. It reduces (halves) the number of chromosomes in gametes.

of 75% with her sisters. Since queens have the normal 50% relatedness to their offspring, a worker who is sterile can do genetically better by looking after a sister (75% of genes in common) than if she were to rear a daughter (50% of genes in common).

Such ideas were suggested by Hamilton (1964, 1972) but, by 1976, Trivers & Hare had pointed out that the increased relatedness to sisters is offset by decreased relatedness to brothers, so helping confers no advantage. They did go on to suggest that this may not be an issue if more sisters than brothers are reared. At this stage, though, they had not considered the value of males compared with females. When this is considered, haplodiploidy confers no advantage on helping yet again. There is a case, noted by Trivers & Hare (1976), where, if the sex ratio in a nest is female biased but that in the general population of nests is 1:1, an advantage to helping may be conferred. In ants this could take the form of some nests producing a preponderance of males and others females. This is the phenomenon of split sex-ratio. Yet even here there are issues, and the current situation is that the haplodiploidy theory may be a red herring and other explanations need to be sought.

Monogamy

monogamy
A situation in which individuals pair for life or the female mates only once.

The notion that monogamy might be important in the evolution of eusociality has been around for a long time, but Boomsma (2009) has suggested that it is, in fact, very important if not crucial. In termites, a king and queen stay together throughout life (what Boomsma refers to as physical lifetime monogamy). In the ants, bees and wasps the queen mates only once in most cases and, although the male dies, his sperm is stored to give a functional lifetime monogamy. Such monogamy can lead to the evolution of eusociality as proposed in Boomsma's monogamy window hypothesis. In this suggestion it is assumed that if a queen is strictly monogamous then her offspring will be equally related to their siblings and their own offspring. Therefore, natural selection will favour situations where it is more efficient to raise siblings than offspring. This could lead to eusociality.

2 General biology of ants

2.1 Nests

Almost all social insects build a nest of some sort. This gives them a home in which to raise young and offers protection against attack and unfavourable weather conditions. The nest also provides a place to store food. In a few species, immobile workers act as food-storage jars, filling their crop with honeydew and staying within the protection of the nest throughout their lives. Other species store seeds as food, and some tropical species establish fungus gardens in the nest.

The most primitive nests are those simply excavated in the soil. Some species that build nests of this kind live mainly underground, feeding on root-sucking aphids (greenfly and blackfly) and soil animals. Their nests are typically horizontal in layout, with chambers often following the roots on which aphids are feeding. *Lasius alienus* shows this pattern. Species that spend time above ground make vertical shafts, with chambers in the soil. *Formica fusca* is one species known to do this. Interesting observations have been made on nest architecture using plaster, latex, solder or molten aluminium poured into the nest (e.g. Tschinkel 2021). These materials set and provide a cast of the internal chamber and gallery structure (Fig. 2.1). However, this technique should be applied only sparingly as it destroys the colony.

Some observations have been made on the behaviour of workers engaged in nest excavation, but we need to know more; this is an area where further contributions could be made, based on observations in the field and in formicaria (see Chapter 7). Most species in Britain that have been studied fall into the simple excavator category, but little is known about the details of nest building in the less common species.

The other main type of nest found in British ants is the mound nest, which may be thatched or unthatched (Sudd 1967). Several species in the genus *Formica* build thatched mounds.

In all ants that excavate, it is natural that the process results in a pile of soil brought up from below ground during construction of the galleries and chambers. Some species carry this soil away from the nest, but others deposit it immediately outside the main point of entry, where it forms a mound. Unstructured mounds of this kind are most common in desert regions; in temperate areas more organised mounds with galleries are built.

Fig. 2.1 Casts of ants' nests made using latex. Top left and bottom left: *Tetramorium caespitum* and *Lasius alienus* (now thought to be *Lasius psammophilus*), both exhibiting a combination of horizontal galleries and vertical shafts. Top right: *Lasius niger*, with just a few horizontal shafts which are sometimes so shallow they break through the surface. Bottom right: *Formica fusca*, showing only vertical shafts. (M.V. Brian)

common names of ants of Britain

Wragge Morley in his 1953 New Naturalist monograph *Ants* suggested common names for 27 species. Vernacular names used here generally follow this. There is a table showing current Latin name, older Latin names and common name where appropriate on pp. 277–278, Table 6.1.

The most striking member of the unthatched mound group is the yellow meadow or hill ant, *Lasius flavus*. The black garden ant, *Lasius niger*, and the large black ant, *Formica fusca*, sometimes build mounds, but these are never as extensive or permanent as those of *Lasius flavus* (Sudd 1967).

These structures are quite fascinating, and often very numerous, so it is worth exploring them further here. Soil is brought from deep down, up to a metre or more, and is therefore often clearly unlike the soil at the surface. Building tends to occur after rain or in the early morning when the vegetation is still wet with dew. The excavated soil is placed onto the wet vegetation around the nest, which allows it to stick. This vegetation helps to support the structure of the nest (Brian 1977). The presence of *L. flavus* nests has consequences for vegetation in the surrounding area. An extensive study of these effects was made by King (1977a,b,c, 1981): his conclusions were that some plant species are confined to the ant mounds, some are unaffected by the presence of the mounds, and some are much more abundant in the surrounding pasture than on the mounds. Some differences in vegetation on nests of *Myrmica sabuleti* and *Myrmica scabrinodis* were observed by Elmes & Wardlaw (1982). Beyond this, little research has been done on this aspect.

The thatched-mound builders par excellence are the wood ants (*Formica rufa*, *F. aquilonia* and *F. lugubris* in Britain, *F. lugubris* and *F. aquilonia* in Ireland, and *F. pratensis* in the Channel Islands). They use a wide variety of long, thin objects to construct their mounds. Leaf stalks, pine needles and small twigs are amongst the materials used (Fig. 2.2). These mounds are not simply accumulations of such materials; they are carefully constructed, and the ants undertake very extensive maintenance work on them (Fig. 2.3).

The nests of the rare *Formica exsecta* are made of soil with a thin layer of twigs and needles on top. They look superficially like the nest of wood ants, albeit small ones (Fig. 2.4). *F. exsecta* is not a wood ant; it belongs to the subgenus *Coptoformica*, all members of which have a notched head. *Formica exsecta* is the only *Coptoformica* species in the region. *Formica sanguinea* occasionally accumulates needles on its nest, but these are nothing like as well organised as those of the wood ants (Fig. 6.11). Finally, *Formica picea* nests could be describe as thatched (Fig. 6.14).

A further type of nest, common in the tropics, is one formed above ground, often in crevices and trees. In Britain, there are crevice-nesting species such as *Leptothorax acervorum* and the *Temnothorax* species. These ants will utilise dead, hollow plant stems. *Temnothorax albipennis* and *Temnothorax nylanderi* are found, for example, in dead stems of viper's bugloss (*Echium vulgare*) at Dungeness, Kent, although both

Fig. 2.2 Wood ants, *Formica rufa*, carrying a small twig to the nest

Fig. 2.3 Thatched nest of *Formica lugubris*

Fig. 2.4 Nest of the narrow-headed ant, *Formica exsecta*, on a roadside verge near Carrbridge, Scotland

Fig. 2.5 Nest of *Temnothorax albipennis* in a dead viper's bugloss stem, Dungeness

Fig. 2.6 A *Lasius brunneus* worker

species can make nests in other situations, such as in acorns and snail shells (Fig. 2.5). They can also be encountered under bark and in the galleries of wood-boring insects when these have been vacated.

Tree nesting, common in the tropics, is poorly represented in the ants of Britain. The brown tree ant, *Lasius brunneus*, is one that does (Fig. 2.6). Their nests are very hard to find, being located deep within the wood of mature, but still living, trees – often oak. The presence of this species can usually be confirmed, however, by the presence of their galleries under the bark (Fig. 2.7).

The other British tree nester is the fascinating jet-black ant, *Lasius fuliginous*. Unlike *L. brunneus*, however, *L. fuliginosus* is unlikely to be encountered in still-living trees but is usually located in old stumps. Here, it constructs a nest from carton, not unlike the material used by wasps in their nests. Carton is a papery substance made of honeydew that the ants mix with small pieces of wood (Maschwitz & Hölldobler 1970) (Fig. 2.8). It is clear that fungi grow in this material and that they help to hold the structure together. What is not so certain is the idea that the ants make use of the fungi as food. Donisthorpe (1915) maintained that *Cladosporium myrmecophilum* fungus was exploited as food, stating that 'it forms a delicate bluish mould on the walls of

Fig. 2.7 *Lasius brunneus* nesting in elm (Peter Furze)

Fig. 2.8 Part of carton nest of *Lasius fuliginosus* (Grzegorz Wagner)

the cells and under the microscope it may be seen to have been bitten off by the ants'. However, this has never been clearly shown to be the case. On the contrary, Maschwitz & Hölldobler (1970) demonstrated that it is not. In the same paper, the authors present a detailed account of how this remarkable ant species makes its nest.

Finally, as noted earlier, some ants do not build a nest at all but live within those of other species, either as inquilines (parasites), xenobionts (guest ants) or lestobionts (thief ants) (see Chapter 3).

Nests help to provide a favourable environment for the colony. The internal environment of a mound nest is remarkably constant. Ants are basically warmth-loving animals and are not generally active below 10°C and cannot produce young below 20°C. Thus, one of the most important things they need to do is to keep a warm nest. There is a significant difference in nest function, and thus in siting and architecture, between tropical and temperate species. Whereas the temperate species nest under logs or stones, or build mounds, those in the tropics more often nest in trees or in rotting logs. For temperate ants, stones form good temperature regulators, warming up in the sun and releasing the heat slowly through the following hours. In spring, a colony close to a stone can get on with such activities as egg-laying much earlier than a colony elsewhere in the soil.

Conditions are not uniform throughout the nest and workers have been shown to move the larvae and pupae around to keep them in the most favourable part of the nest at all times of day. This has been well worked out for a Japanese species, *Formica japonica* (Kondoh 1968). This is research that could profitably be repeated with some of the species in Britain. Nest climate (or, more correctly, microclimate) is more closely regulated in those species that build mound nests than in other ants.

As long ago as 1810, Huber suggested that the main function of mound nests was to regulate microclimate. That notion has been supported by extensive investigations since then. The most obvious microclimatic variable that might be regulated is nest temperature. The temperature 30 cm or so below the surface of a nest mound of thatch-building *Formica* species varies very little (Fig. 2.9). It is much more

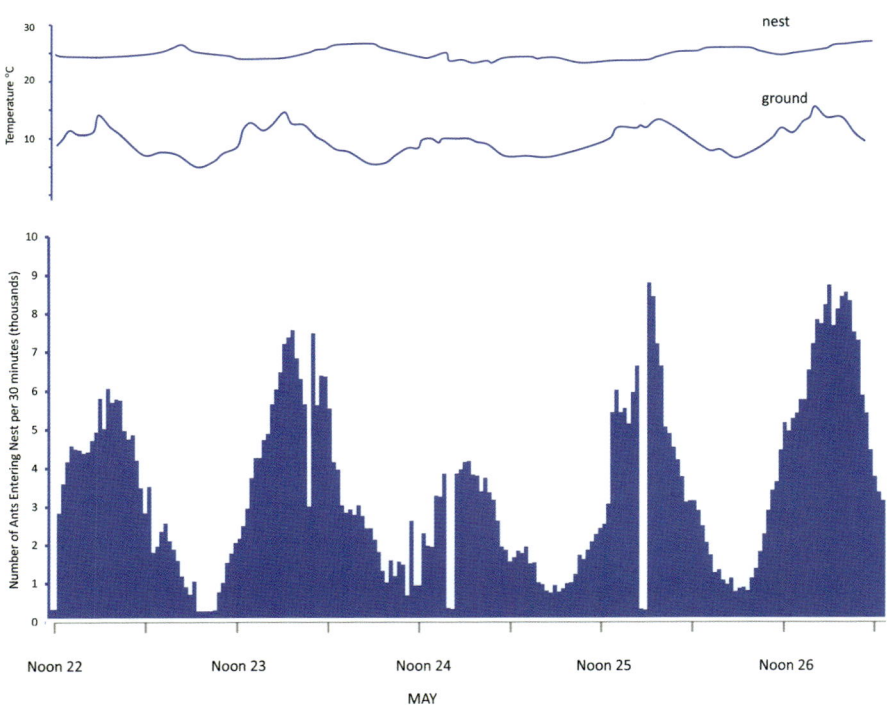

Fig. 2.9 Nest temperature and ant activity in a wood ant (*Formica rufa*) nest over a four-day period (from Skinner 1980a)

constant than that in the surrounding soil, and it is close to that which the ants prefer. Wood ants have the same temperature requirements as tropical species but live in much colder environments. This is possible because of the nest architecture. But how is the temperature regulated in this way? There seem to be two major factors involved: insulation and aspect. Insulation, slowing down heat loss, is achieved by thatching the outer crust of the mound, and maintaining the thatch in the face of erosion by wind and rain. Air trapped in the mound may also help to insulate the nest. Interception of the sun's radiation depends on aspect, as it does in a solar panel. The south-facing slope of the nest is shallower than the others, so that it intercepts the sun's rays at solar noon at right angles, thereby maximising the amount of warmth.

Yet the sun is probably not the only source of heat for the nest. The nest may be warmed by heat from the ants themselves, along with that produced by the decay of organic material, of which there is always much in the mound. There is some controversy over this latter point. Some researchers believe that decomposition makes a significant contribution to heat input, whereas others consider that the heat produced by the ants themselves is much more important. Work on *Formica polyctena* by Horstmann & Schmidt (1986) implies that worker behaviour may have a very important part to play in regulation of temperature. Zahn (1958) goes as far as to suggest that the spring massing behaviour seen in *Formica* species actually functions to carry heat into the nest, within the bodies of workers that have been sunning themselves (Fig. 2.10). This interpretation is not accepted by all researchers (e.g. Kadochová *et al.* 2017). However, there is little doubt that the workers do contribute to the regulation of nest temperature less directly by constantly repairing the thatching of the mound. In several experiments in which nests were destroyed or disturbed, there was immediate and widespread worker activity to put things right. Evidently, the mound is not simply an unstructured accumulation of soil and fragments of vegetation. Schwenke (1957, quoted in Sudd 1967) found that up to half of all loads brought to the nest in the month of June were plant material, indicating the importance of nest integrity.

There is much scope for further investigation in this area, particularly to see whether conclusions based on studies elsewhere also apply to species in Britain. There has been much confusion in the past about the taxonomy of European species of *Formica* and thus information from continental

Fig. 2.10 Spring mass of wood ants. *Formica lugubris* at Hebden Bridge, West Yorkshire, 25 February

studies must be viewed with caution, especially if they are rather old. Much of this work would bear careful repetition and extension on critically identified species in Britain, now that the taxonomy of our species is clearer.

Ants can regulate the humidity as well as the temperature of the nest. Severe desiccation will kill all ants, but experiments demonstrate that some species are more resistant than others. Ants placed in a dry container will eventually die, but tree-living species take much longer to do so than those from other habitats. In an experimentally created humidity gradient, workers have been shown to move the larvae and pupae into a preferred region where the soil moisture is similar to that found within the nest. For example, *Formica ulkei* selected a soil moisture content of about 30% in this way (Scherba 1959). The moisture content of the nests that Scherba monitored were close to 30%, and the range in the nest was lower than that in the surrounding soil. It is not clear how this regulation is achieved, but it would seem to involve nest structure. In some species living in very dry habitats, workers have been seen collecting water in various ways. It would be worthwhile for these investigations to be repeated on other species in Britain.

Generally, ants in Britain do not encounter severe dehydration problems. For instance, only after very hot, dry spells does the soil in the upper part of a *Lasius flavus* mound become less than saturated with moisture (Brian 1977). When this happens, the ants move downwards into cooler, more humid regions below.

Just as a person's home is their castle, so is the nest of ants, and it may well have to be defended. What is more, the defence often extends beyond the boundary of the nest itself into an area around it. This area, which is patrolled and defended, is the territory or home range. Many animal species attack nests, including insectivorous birds such as the green woodpecker, *Picus viridis* (Figs 2.11, 2.12). The number of ants eaten can be estimated by counting heads in the droppings. A study of this predation on wood ants in Lancashire gave a figure of about 3,000 ants eaten from each nest over a winter. This is less than 1% of the estimated nest population (Skinner 1976). It is well known that many game birds eat ants, and it was common practice in the past to introduce wood ants into woodland as food for pheasants, *Phasianus colchicus*. The wryneck, *Jynx torquilla*, also feeds on ants, but it is too rare to be an important predator. There is also evidence for attacks by badgers, *Meles meles*.

myrmecophile
Any organism that relies on ants for at least part of its lifecycle.

Ants are also attacked by some of the species that live with them (myrmecophiles). Such 'guests' are called *hostile mymecophiles* (Stockan *et al.* 2016). See Chapter 3 for more on this.

The most likely attackers are other ants of the same or different species. This risk is countered by a sophisticated

Fig. 2.11 Droppings of green woodpecker on a wood ant (*Formica rufa*) nest. These droppings may contain the remains of up to 100 workers.

Fig. 2.12 Detail of material found in one dropping. Notice the heads, which are arrowed.

recognition system. It has long been believed that each species, or even each colony, has a distinct odour. But only in recent times has it been possible to identify the complex mixture of chemicals involved. Because the quantities involved are so tiny, it was only with the advent of techniques such as gas–liquid chromatography that these odours could be characterised. The idea that ants produce species-specific odours is well established, but less is known about the basis of colony specificity that enables individuals to recognise friend or foe of the same species. Perhaps colony odour is derived from the environment and food of the colony, or perhaps a specific mix of volatile chemicals are inherited from the queen. D'Ettore & Lenoir (2010) distinguish between kin and nestmate recognition. If the colony exhibits polyandry (where the queen mates with more than one male) or polygyny (where there is more than one queen), then nestmates may not actually be full siblings. Although kin selection may have been a major factor in evolution of eusociality (p. 10), it is largely replaced in modern ants by what Lenoir *et al.* (1999), call 'nestmateship'. Fascinatingly, some experimental studies have shown that different species can be induced to recognise each other as nestmates or fellows (e.g. Jaisson 1991). Errard (1994), for example, created mixed colonies of *Manica rubida* and *Formica selysi* (neither is found in Britain or Ireland). This kind of study could bear repetition and extension on species in Britain. Chapter 7 has details of ant culture methods.

The question arises: what are the cues by which kin, nestmates or fellows recognise each other? The most widely accepted suggestion is they are mainly found in the cuticular hydrocarbons (CHCs) which cover the outer 'skin', or cuticle, of all insects. This waxy layer provides a waterproof outer coating and probably evolved initially for this reason. Analysis of the CHCs of ants reveals that they differ between species in terms of which CHCs are there and vary in amount within species. Through food sharing (trophallaxis, p. 39) and touching (allogrooming), these CHCs become uniformly distributed through the colony to give the colony odour.

Studies by Guerrieri *et al.* (2009) have shown that carpenter ants do not specifically recognise nestmates but do spot and reject non-nestmates which have odours novel to their own colony. They go on to say, 'this begs for a reappraisal of the mechanisms underlying recognition systems in social insects'. Science rarely rests, however, and in a more recent paper Martin *et al.* (2019) suggest that, at least in *F. exsecta*, the workers must be aware of the differences between their

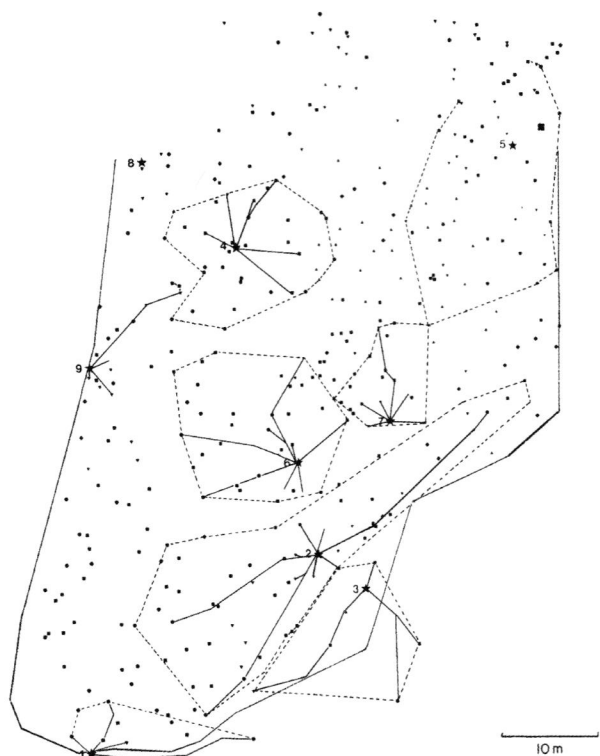

Territory and trail patterns, Cringlebarrow Wood. Summer 1973. ■ Sycamore, ● Oak, ▲ Yew, ▼ Scots Pine, ♦ Ash, ★ nest of *F. rufa*.

Fig. 2.13 Map of nests and their territorial boundaries (dashed lines) and trails (solid lines) of the wood ant, *Formica rufa*, in Cringlebarrow Wood, Lancashire, 1974 (from Skinner 1980a)

own and their nestmates' odour profile. This is clearly an area in which much remains to be learned.

This is a complex and fast-growing area of research and readers are directed to D'Ettore & Lenoir (2010), Morgan (2008) and Martin (2016) for excellent, accessible and up-to-date reviews of this subject. Hoyt (1998) gives a very readable account of the many kinds and importance of ant odours based on pheromones.

How efficient is this recognition? Invasion of a nest by a different species of ant elicits a violent and almost instantaneous attack from the occupants. If the intruder is of the same species, but a different colony, the attack may be just as violent, or it may simply involve offering less food to the newcomer, until it has acquired the colony odour. In most species, the response is so distinct that it can be used to discover colony affinities of individuals. Workers of unknown affinity are placed on a test nest and the response of resident workers is compared with their response to 'control' workers from their own colony that are removed and then replaced. These observations give a clear answer to the question 'friend or foe?' This simple technique can be used to map territorial boundaries with some accuracy (Fig. 2.13). The method is described further on p. 294.

2.2 Reproduction

All ants have the same basic lifecycle. Generally, eggs are laid only by queens. They hatch to give a legless larva which is fed by workers. It moults, usually three times, and finally pupates. In this inactive pupa (in some species encased in a silken cocoon and in others not), the larval tissues are completely reorganised to form the adult (imago) (Figs 2.14–2.17). This will emerge from the cocoon in a pale, callow state. Recently emerged workers are recognised by their pale colour, but they darken within a few days (Fig. 2.18). The specialised females, called queens, are generally the only ones that lay eggs. Even if workers do lay eggs, these are nearly always trophic eggs used as food. Occasionally, worker eggs hatch. These produce males, because in the Hymenoptera males develop from unfertilised (haploid) eggs laid by queens or workers. Fertilised eggs, with the potential to become new queens or, more often, workers, can be laid only by queens,

callow
A newly emerged ant.

trophic
Related to feeding and nutrition.

Fig. 2.14 Larvae of *Tetramorium caespitum*

Fig. 2.15 Worker cocoons of *Lasius* species wrapped in silk spun by the larva

Fig. 2.16 Gyne (right) and worker (left) cocoons of *Lasius niger*

Fig. 2.17 Pupa of *Myrmica* species (bottom left). Note there is no silken cocoon in this subfamily of ants.

Fig. 2.18 A callow worker of *Lasius flavus*

because only queens mate. Colonies produce the sexual forms (that is, winged, unmated females and males) at various times of year, depending on species (Figs 2.19, 2.20). In the wood ants, sexual forms appear in May, having been reared on food stored over the winter. In most other species, sexual forms appear in midsummer (Table 2.1).

Generally, all the nests of the same species in a wide area will release their winged forms on the same day, which is usually a warm and sultry one. Millions of these 'flying ants'

Fig. 2.20 Winged female (queen) of *Formica pratensis*

Fig. 2.19 Winged male of *Formica rufa*

may emerge on such a day. Most will not live to reproduce, as they will be eaten by insectivorous birds and by other insects. There is much still to be discovered about the pattern of nuptial flights. Most seem to fall into one of two broad types, which Hölldobler & Bartz (1985) refer to as the female calling syndrome and the male aggregation syndrome. In the former, wingless females remain on the ground near the nest, releasing sex pheromones, 'calling' winged males to them. This pattern is generally found in more primitive species with small colonies and a low rate of production of sexual forms. In such species, synchronisation of flights in an area is less marked than in species displaying the male aggregation syndrome. As in any attempt to codify complex animal behaviour, this dichotomy is a simplification, albeit a useful one. Peeters & Molet (2010) present a table suggesting subdivisions of female calling and male aggregation strategies. As an example, in *Lasius* species, males fly at a fixed range of heights above the ground. The females enter the swarm of males. They then mate and disperse widely before shedding their wings and excavating a nest.

Most of our species are of the latter type, male aggregation, but information is fragmentary so this is an area where much could be achieved. A very useful study pointing the way to possibilities is that of Brian & Brian (1955).

In the male aggregation syndrome, the males congregate in various prominent places such as the tops of hills, trees or buildings, in large swarms made up of males from many colonies. The females fly into these swarms, where they mate

Table 2.1 Some of the features of the nests and reproductive biology of ants in Britain, compiled from a variety of sources including Seifert (2018), Lebas *et al.* (2016), Brian (1977) and personal observations. It summarises current thinking on nests and reproductive biology for the species found in Britain.

Species	Nest type and locality	Gyny*	Domy**	Founding	Nuptial flight
Tapinoma erraticum	under stones or a mound of debris, shallow often associated with a hot metal object such as a can in Britain with cooler summers	poly		either adoption or claustral	early summer
Formica exsecta	small soil domes topped with plant debris	mono or poly	mono or poly	temporary social parasitism on *F. fusca, F. lemani* and *F. cunicularia*	mid–late summer
Formica lugubris, F. rufa, F. aquilonia	domes of pine needles up to 150 cm	mono or poly	mono or poly	temporary social parasitism on *F. fusca* or *lemani* or adoption	early–midsummer
Formica sanguinea	in old rotting tree stumps or under stones, may be covered in twigs	mono or poly		temporary social parasitism on *F. fusca* or *lemani*, adoption or single queen	midsummer
Formica cunicularia, F. fusca, F. lemani, F. rufibarbis	under stones etc., in rotten wood	mono or poly		independent by single queen, claustral	early–midsummer
F. picea	in sphagnum or other moss, often with a cone of plant material on top	poly only		independent by single queen or adoption	mid–late summer
Lasius flavus	mound of earth covered in vegetation up to about 30 cm	mono or poly		independent by several queens (pleometrosis)	late summer–early autumn
Lasius umbratus, L. meridionalis, L. mixtus, L. sabularum	underground tunnels, in roots and dead wood, walls made of a papery substance			temporary social parasitism on other *Lasius* species	spring–autumn
Lasius fuliginosus	in hollow trees, made of papery carton	mono, sometimes poly		temporary social parasitism on members of the *Lasius umbratus* group. This makes *L. fuliginosus* the only hyperparasitic*** ant in Europe, since its hosts found by parasitism too.	late spring–early autumn
Lasius alienus, L. psammophilus	under stones or in ground with a mound of earth above	poly			mid–late summer
Lasius brunneus	in trees, but can be found in ground and walls	mono		claustral	early summer
Lasius emarginatus	anywhere there is a suitable cavity	mono		claustral	midsummer
Lasius niger	a small mound of soil and under stones etc.	mono		independent by several queens	midsummer
Lasius platythorax	in stumps and branches, no earth mound	mono		independent by several queens	midsummer
Formicoxenus nitidulus	in domes of red wood ants	poly		independent and budding	midsummer

Leptothorax acervorum	in dead wood or under bark	mono or poly	independent and budding in polygynous colonies	midsummer–early autumn
Myrmecina graminicola	under stones, moss etc.	mono, occasionally poly	independent	mid-spring midsummer
Myrmica lobicornis, M. sulcinodis	under stones, logs	mono	independent	late summer
Myrmica rubra	under stones etc.	poly	budding and adoption	midsummer
Myrmica ruginodis	under stones, in grass tussocks	mono or poly	sometimes poly independent	midsummer
Myrmica sabuleti	under stones	poly	independent by single queen or adoption	late summer–early autumn
Myrmica scabrinodis, M. vandeli	under stones, in dead wood, under bark	mono or poly	independent by single queen or adoption	midsummer–early autumn
Myrmica schencki	under stones, in ground, entrance often has a small chimney of plant material	mono or poly		midsummer–early autumn
Myrmica specioides	under stones	mono or poly	independent by single queen or adoption	late summer–early autumn
Myrmica hirsuta, M. karavajevi	parasitic in nests of other ants of Myrmica genus			late summer–early autumn
Stenamma westwoodii, S. debile	under stones, in moss	mono or poly	claustral, adoption	early–mid-autumn
Strongylognathus testaceus and Tetramorium atratulum	Parasites in nests of Tetramorium caespitum			midsummer
Temnothorax interruptus	under stones, in ground	poly	claustral	midsummer
Temnothorax albipennis, T. nylanderi and T. unifasciatus	in dead plant stalks, acorns, snail shells etc.	mono	independent, intraspecific parasitism	midsummer
Tetramorium caespitum	under stones, in ground, often quite deep	mono	independent	early summer–mid-autumn
Ponera coarctata, P. testacea	deep in the ground	mono or poly		midsummer

*   Gyny has two states: monogyny where the colony has just one queen, or polygyny where the colony has two or more queens.

**   Domy has two states: monodomy where the colony is housed in one nest, or polydomy where the colony is housed in two or more separate nests.

***   Hyperparasite: a parasite that uses a host that is itself a parasite.

and then disperse, to eventually shed their wings and begin founding a new nest, if they survive long enough (Fig. 2.21). A successfully inseminated queen sheds her wings when these have served their function of dispersal. The muscles that operate them are to be used as a source of food in the days ahead. In many species, the queen alone must feed the first batch of larvae that hatch from her first-laid eggs. Material derived from the breakdown of the wing muscles contributes to the food for the larvae as well as to creating the eggs themselves. In some ant species the queen builds a small nest and leaves it periodically to find food for her brood. This 'partially claustral' pattern is thought to be the primitive state. In more advanced, 'claustral' forms the queen constructs a small nest chamber and seals herself into it to live on food stores in her body until the first workers emerge. Certain ant species found colonies differently. In the wood ant *Formica rufa*, for example, mated queens sometimes return to the home nest (known as adoption) and then possibly move to a new site with some workers, which is known as budding.

Another pattern emerges when mated queens of, for instance, *Formica lugubris* fail to get back to their own nest. They may then invade the nest of certain other species, such as *Formica fusca*. When this happens, the *F. lugubris* queen eliminates the *F. fusca* queen and takes on the role

claustral

A form of behaviour in which the fertilised queen ant seals herself in a chamber and stays there rearing the first brood on her own.

Fig. 2.21 Dealate (having lost the wings) female of *Formica pratensis*

of queen to the *F. fusca* workers. She also lays eggs. These, and the larvae that emerge from them, will be tended by the *F. fusca* workers. Eventually all the *F. fusca* die off, and because they are not replaced, they leave a pure *F. lugubris* colony (Buschinger 1968). This method, known as temporary social parasitism, is probably less common than the budding described above. *F. rufa* is a more regular temporary social parasite, and for this reason most of the colonies are single-queened (monogynous, although note that the F. *rufa* × *F. polyctena* hybrid now known to occur in Britain is probably polygynous). The hosts are *Formica fusca* and *F. lemani*. The approach of the *F. rufa* queen is not very subtle: she simply barges into the nest. She often perishes here at the hands of the workers of the other species, but enough attempts are successful to make this a fairly common species, although it is currently declining due to loss of habitat in Britain. It has not been found in Ireland.

The rare *Formica exsecta* founds its colonies in a similar way, but the small shiny *F. exsecta* queens have a more subtle approach to colony penetration. Often, they are carried into the nest by workers or enter in a stealthy manner. They do not elicit such hostile behaviour as other queens. Colony founding is similar in *Lasius fuliginosus* and *Lasius umbratus* and other members of these two species groups. The host is any member of the *L. niger* group of species. In this case, however, the relationship is obligatory; the parasite species have no alternative method of colony founding. Hölldobler (1953) showed that a queen of *L. umbratus* first kills a worker and runs around with it in her mandibles before attempting nest entry. The host queen is disposed of. It is not known how this is achieved, but in the Austrian *Lasius reginae* the host queen (*Lasius alienus*) is 'rolled over and throttled' according to Hölldobler & Wilson (1990) quoting Faber (1967). Seifert (2018) elaborates on what happens and suggests that the ventral nerve is cut by a sawing movement of the dents on the mandibles.

In non-parasitic species, which found colonies 'from scratch', the first workers to emerge (known as nanitics) are rather small and timid compared with those found in a mature colony. These features seem to be adaptive, in the sense that there is a balance between the number of workers and their size. The queen has limited resources and she can use these to produce a few large workers, or many smaller ones. Many workers will be able to do all the jobs of food finding, nest enlargement and so on, but if the workers are too small, they will not be able to exploit the normal food

species group
Many ant genera are divided into subgenera which form species groups. *Lasius fuliginosus* is in the subgenus *Dendrolasius* (the only one in this subgenus in Europe), *L. umbratus* is in the subgenus *Chthonolosius* (along with 11 other species in Europe, three of which occur in Britain). Lebas *et al.* (2016) have more details on subgenera.

sigmoid growth curve
An S-shaped curve,
showing a slow start, a
rapid rise and, finally,
a levelling off.

and nest sites of the species. It should therefore be possible to predict the optimum number of newly emerged workers, and this predicted number of workers has been shown to be close to the observed number in at least one species, *Solenopsis wagneri* (=*Solenopsis invicta*).

After a hazardous early stage, the colony, if it survives, will enter a period of sustained exponential growth. In those species that have been studied, the growth curves show a typical sigmoid form, but it is also clear that the factors controlling the growth of the population are much more complex in ants than they are in the non-social insect species. Eventually, before it reaches its maximum worker population, a colony begins to reproduce. Much work has been done to identify the factors controlling the onset of this reproductive phase in a colony's history. Brian and his team at Furzebrook in Dorset studied the colony history of species in the genus *Myrmica*. In one series of observations, Brian was able to show that colonies often produce males before any new queens appear, and he described a colony in this male-producing phase as 'adolescent', as opposed to the earlier 'juvenile' phase and the later 'mature' phase in which queens are produced as well. Brian also demonstrated that the colonies enter the adolescent and mature phases due to a dilution of queen influence. In *Myrmica ruginodis*, the switch occurs when there are 900 to 3,000 workers, but the crucial figure is the ratio of workers to queens, and in this species the relevant ratio is 1,000 workers per queen. It has been shown that the queen dilution effect arises when she no longer touches each of the workers often enough to have proper control over their behaviour towards the brood. This influence seems to act via a chemical, because dead queens are effective when intact but not when their lipids have been extracted. Brian's summary of the situation is illustrated in Table 2.2.

In other species, it has been shown that workers bite repeatedly at the larvae, and this causes the larvae to develop into workers. But, apart from this work on *Myrmica* species, little is known about the timing of the change from juvenile to adolescent and mature stages. Thus, it is not known whether the same pattern is found in other species. Nonacs (1986) reported that some colonies will produce only males in a particular season, whilst others produce only queens and yet others a mixture. It is not known whether the strategy of an individual colony changes from year to year.

Table 2.2 A summary of influences that determine caste in ants (after Brian 1977)

	Behavioural Type	
	Queens present	Queens absent
1. Worker-biased larvae	Nurses feed these actively in preference to other larvae and they metamorphose into workers.	These larvae are neglected, and they stop growing part-way through the third instar.
2. Labile female larvae	Workers (mainly foragers) attack these larvae during the stage in spring when they are signalling their ability to become gynes by means of a cuticular secretion.	Both workers and foragers feed these larvae protein-rich foods focusing on a few at a time.
3. Male larvae	Neglected but not attacked.	Fed in preference to worker-biased larvae but not labile female larvae that are forming gynes.
4. Egg formation and laying	As long as the queen is laying or there is a cluster of reproductive eggs of either sex, young nurse workers lay imperfect eggs that are useful only as food (trophic eggs). Very young workers (<three weeks old) will lay reproductive eggs that form before they become sensitive to queens.	Young workers lay large perfect reproductive eggs which are unfertilised but capable of developing into males parthenogenetically.

gyne
The reproductive female caste of ants, effectively queens.

caste
In ants the queens and workers, as two 'types' of female. Males are not a caste, although many researchers and others treat them as if they were.

gaster
From the Greek for stomach. In Hymenoptera, it is the rear part of the abdomen behind the petiole (see Fig. 6.3).

2.3 Caste and division of labour

A colony may contain ants that look and behave very differently from one another. At the simplest level, the colony holds three types of individual: males, which are generally short-lived and do no work and are not technically a caste; the queen caste, which founds the colony and lays eggs; and the worker caste, which performs all the other tasks. Workers do not generally reproduce. Many workers lay eggs, but these are usually used as food. Less often, they give rise to males. In addition to males, queens and workers, various other forms are recognised. Ergatogynes are reproductive forms morphologically intermediate between queens and workers. In some cases, these replace the queen. Such forms are found in the British species *Formicoxenus nitidulus* and *Temnothorax nylanderi*; neither are known to occur in Ireland. Forms not known in any species in Britain include the gamergates (reproductives with the morphology of workers) and dichthadiiform ergatogynes (queens with abnormally enlarged gasters). Chapter 6 gives details of the morphological differences between workers, queens and males.

Many species exhibit worker polymorphism where the worker caste is further divided into morphological subcastes. The major (large) workers are called soldiers when their only job is fighting. Minor (small) workers are the best-known forms, but media (middle-sized) workers are also recognised in some species. This morphological segregation of workers

Fig. 2.22 Polymorphism in the ant *Pheidole tepicana*, showing major, media and minor workers (Alex Wild)

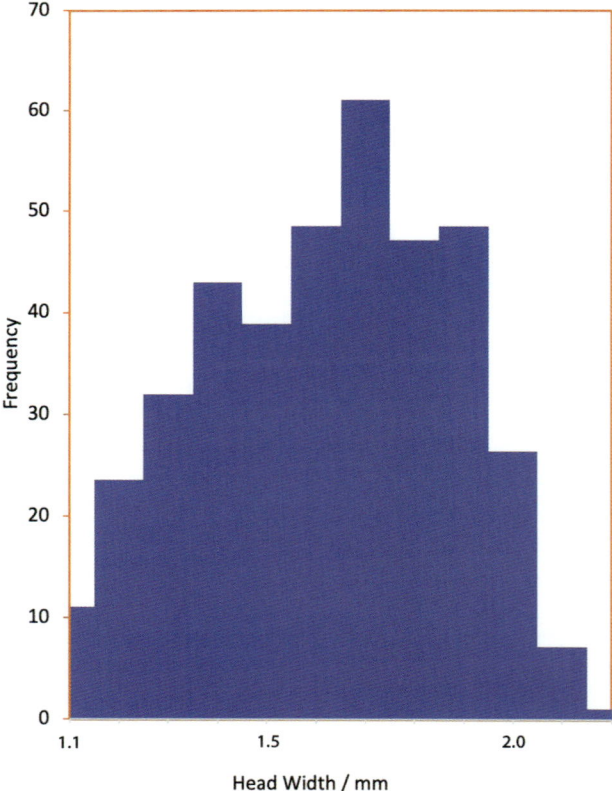

Fig. 2.23 Frequency histogram of head widths in *Formica rufa* (from Skinner 1976)

into subcastes is almost always directly related to a division of labour. These subcastes are not found in any of the ants of mainland Britain or Ireland, but that does not mean that the workers are all alike in appearance or in behaviour. Size variation in the workers of our species is continuous, not discontinuous as in species with definite subcastes. For example, in the myrmicine *Pheidole tepicana*, which is not found in Britain or Ireland, the major, media and minor workers are clearly distinguishable in terms of head width (Fig. 2.22), whereas in *Formica rufa*, head-width measurements show variation but no obviously distinct forms (Fig. 2.23).

myrmicine
A member of the subfamily Myrmicinae.

Whether or not there are clear worker subcastes, workers tend to change their jobs as they grow older. This phenomenon, known as age-specific polyethism, or temporal castes, is seen in all other social insects too, and is well documented in honey bees (Gould & Gould 1988, Seeley 1985).

For the first few days after emerging from the pupa, young workers generally remain in the nest, tending brood and the queen. Later they begin to make foraging trips from the nest and become bolder in the investigation of foreign objects. Amongst the ants of Britain, this change of job with age has been shown in some species (Table 2.3).

It would be useful to gather data for such species as *Formica fusca/lemani*, other *Myrmica* species and *Lasius fuliginosus*. Such studies are probably most easily carried out on laboratory colonies (Chapter 7) but could also be done in the field. Some ants darken as they grow older, so it is possible to estimate the age of individuals, even if they have not been experimentally marked.

In a continental species, *Formica polyctena*, Dobrzanska (1959) investigated age polyethism by marking individuals with rings around the waist. Workers did not begin to forage until 45 days after emergence from the pupa. For

Table 2.3. Species in Britain in which age polyethism has been demonstrated

Species	Reference
Myrmica rubra	Ehrhardt 1931
Myrmica scabrinodis	Buschinger 1968
Leptothorax acervorum	Weir 1958a,b
Lasius niger	Heyde 1924
Formica sanguinea	Dobrzanska 1959, Billen 1984
species in the wood ant group	e.g. Otto 1958, Rosengren 1987

the first few days they moved very little. At three days they began to take food from nestmates. At about six days they began tending brood. Then followed other jobs such as removing eggs, regurgitating food to young larvae and giving older ones bits of food. At about 27 days the workers began to exhibit aggressive behaviour and started spraying formic (methanoic) acid. It is worth noting that it is now almost certain that British *Formica rufa* are in fact hybrids of *F. polyctena* and *F. rufa* (Monaghan 2022), so it is likely that these findings of Dobrzanska are relevant to British *F. rufa*, but confirmation and extension of these studies to other species would be useful. *F. rufa* does not occur in Ireland.

Associated with these behavioural changes are physiological ones. Brood-caring workers have well-developed ovaries, but the ovaries of foragers caught outside the nest have degenerated.

Within these broad categories of tasks, called by German myrmecologists *Innendienst* (inside work) and *Aussendienst* (outside work), individual worker ants specialise in certain jobs. Again, this can be investigated by marking experiments. Otto (1958) revealed that in wood ants the smaller outside workers tended to specialise in honeydew collection.

In the ants, as in most, but not all, bees and wasps, males arise from eggs that are unfertilised and therefore haploid. On the other hand, females (queens and workers) arise from fertilised eggs which are diploid (having two sets of chromosomes). Amongst females, caste is determined by physiology and not by genetics. That is, all fertilised eggs have the potential to become either workers or queens; whether a particular fertilised egg develops into a worker or a queen does not depend in any way on its genetic make-up. Several other factors seem to be important. The presence of a queen is one such factor. In colonies from which queens were experimentally removed after eggs had been laid, more larvae developed into queens than would otherwise have done (Brian & Carr 1960). In *Formica polyctena*, the proximity of queens has been shown to be crucial. Queens of this species lay their early-season eggs in the top of the nest, and then move down into the lower chambers. These early-season eggs have a strong tendency to develop into new queens. In experiments in which existing queens were prevented from moving away, these eggs failed to develop into new queens but became workers instead. The situation is similar in a related species, *F. pratensis*, but in this case the queen-suppressing effect of the presence of an existing queen can be overridden by a very large

haploid
Having only a single set of chromosomes.

chromosome
A long molecule of the genetic material DNA, folded in a protein scaffold. Ants generally have more than one pair of chromosomes in the diploid state. However, the Australian jack jumper ant (*Myrmecia pilosula*) has only one pair, which means in the male (which is haploid) there is just one chromosome, which is the lowest number possible.

worker–queen ratio of 600:1 (Gosswald & Bier 1953). The presence of a queen also seems to inhibit worker reproduction. In some species, more worker eggs were laid in colonies without queens than in colonies with queens. For example, 50 *Temnothorax* (formerly *Leptothorax*) *unifasciatus* workers laid 140 eggs when queens were absent, but only 60 in the presence of queens (Bier 1954). It is not clear how queens exert this influence over brood development. In honey bees it is well known that the queen exerts a similar influence via a chemical message, or pheromone, called queen substance, and this has been identified as (*E*)-9-oxo-2-decenoic acid. Does such a controlling pheromone exist in ants? Work by Holman *et al.* (2010) suggests that it does, at least in *Lasius niger*.

Carr (1962) demonstrated that dead queens, if regularly replaced, inhibited the growth of larvae, and later it was shown that this inhibiting effect was due to a fatty acid that could be extracted from the heads of queens. More recent work has implicated other factors as well. Firstly, cytoplasmic factors are involved. Eggs do vary in their ability to become queens or workers, perhaps because of differing amounts of reserves in the cytoplasm. Young queens tend to lay worker-biased eggs, and older queens lay more queen-biased eggs. A third factor of importance seems to be temperature. If it exceeds about 20°C in the last larval stage, workers tend to develop, since queens do not develop at high temperatures. Brian (1977) suggested that *Myrmica* is adapted to, or perhaps even a prisoner in, cold climates. The path of development – to worker or queen – remains undetermined and susceptible to environmental modification through much of the larval and pupal stage. The quality of the food supply is important. A larva in a well-established colony will grow well and hibernate over the winter. If the food supply is still good in spring, the larva will continue towards queenhood, but if not, it will develop quickly into a worker. Queens influence the path of brood development through their effect on worker behaviour. In the absence of queens, workers give extra food to the larger larvae, which are those most likely to become queens. When queens are present, larger larvae receive less food, and both workers and queens may attack them, either killing them or inducing early metamorphosis into workers.

Hölldobler & Wilson (1990) summarised this complex picture very nicely: 'six factors … determine whether a *Myrmica* female will become a worker or a queen: larval nutrition, winter chilling, post-hibernation temperature,

cytoplasm
The jellylike material that fills the cell. Embedded in the cytoplasm are organelles, such as mitochondria and the nucleus. These organelles carry out specific functions, such as respiration in the mitochondria, and the storage of the genetic material in the nucleus.

queen influence, egg size, and queen age'. The next question should logically be, what is their relative importance in nature? A useful way to view the entire caste-determining system is to regard it, metaphorically, as a series of checkpoints arranged more or less in sequence. An egg 'aspires' to develop into an adult queen. This ambition is 'approved' by the colony, providing the following two checkpoints are passed. First, has the larva been through diapause and chilled before resuming development? Second, has the larva reached the requisite size by the onset of adult development in the final larval instar? In addition, are the mother queens nearby and potent, and is the colony young? If so, borderline cases are more likely to fail the queen test and be consigned to workerhood. Taken together, the caste-biasing factors make it more likely that the *Myrmica* colony will produce new queens in the spring and also when it is large and robust, the conditions under which it can most profitably invest in reproduction.

This complex situation has been worked out in species of the genus *Myrmica*, but what of other genera? The other species group in which detailed studies have been carried out is the *Formica rufa* group, and this shows several differences from *Myrmica*. The first major difference is that the ants hibernate without any brood. Queens move to the upper parts of the nest when the temperature rises in the spring and begin laying eggs. At first the temperature is too low for fertilisation to occur and so these very early 'winter eggs' become males. When the temperature rises above about 19°C, the next batch of winter eggs laid are fertilised and can become workers or queens. Some *Formica* species can regulate nest temperature, but small colonies do this less effectively than larger ones, and thus tend to produce more males. The eggs laid in summer all become workers. The final destiny of a later, winter egg – worker or queen – seems to be determined in the first three days of larval life. Again, the presence of queens is important. When none are present these winter eggs always develop into queens. As we have seen, in a natural colony the larvae are effectively queen-free at this stage, because the queens have gone deep into the nest. Recently hibernated workers are very active and have well-developed nutrient-producing glands, and this increases the tendency of winter eggs to develop into queens. Thus, the system is similar in some respects to that found in *Myrmica* species: eggs are biased towards worker or queen; there is a seasonal trend in this bias; queens suppress queen formation; workers vary with season in their ability to nurse;

and food availability and temperature are important. Most other studies have shown only minor variations on these basic patterns, but there are some very different situations, and further study of other species would be worthwhile.

2.4 Feeding

Ants spend much of their time collecting food. Worker ants are exclusively responsible for this job within the colony. Males rarely feed, and the queens and the young stages are provided with food by the workers. Most other animals collect food for themselves alone, or possibly for their own young, but much of the food collected by an individual of a social species is not for itself at all (Fig. 2.24). A consequence of this is that food must be carried back to the brood and queens in the nest. It is therefore possible to monitor the food intake of ant colonies. In most other invertebrate animals, specialised techniques are required to monitor food intake,

Fig. 2.24 Worker ants, *Formica rufa*, feeding each other, a phenomenon called trophallaxis (p. 23)

but with some species of ants much can be learnt from simple observational studies. Most of this work has been done on species that form obvious trails, especially the wood ants and others such as *Lasius fuliginosus*. Such studies reveal that ants subsist on a wide range of food items. *Formica rufa* has been shown to carry many types of prey, as is the case with other species that have been investigated (Table 2.4).

Such a list can be produced quite quickly by observing trails leading to the nest, but it may take rather longer to get an idea of the quantities of material brought back (Fig. 2.25). Brief summaries of the food of every species found in Britain can be found on the BWARS website. In truth, though, the information is often preliminary and quite speculative. It is clear that much could be added to our knowledge of the food of the less-known of our species by the diligent and patient observer.

In the study of wood ants described above, five-minute periods of observation were long enough to give useful information, but more sustained observation periods would be needed if there were less traffic of food items along the trail. To quantify food intake of a colony, it may be necessary to make observations over periods of an hour or more. The importance of each type of food for the colony is not shown by numbers alone. For example, a caterpillar brought back by wood ants may be many times larger than a single aphid. A single observer can look at only one trail at a time. Although it may be possible to get some idea of what is being carried back by simply looking, very often it is necessary to confiscate the food item for closer inspection. This inevitably disturbs the ants. The deprived worker will release chemical alarm signals, which affects other foragers on the trail. It is rarely possible to study an animal without such an 'observer

Table 2.4 Items carried back to nests by ant species found in Britain (*F. rufa* from Skinner personal observations, others from Brian 1977).

Species	Food observed
Formica rufa	aphids, bark lice (Psocoptera), beetles, earthworms (usually in bits), bumblebees, a variety of small flies, especially St Mark's flies (*Bibio* species), woodlice, moth caterpillars (especially winter moth *Operophtera brumata*), small spiders
Myrmica species	aphids, springtails, larval and adult flies, spiders, other ants
Lasius flavus	soft-bodied mites, beetle larvae (mainly wireworms of two species), woodlice, *L. flavus* workers and queens
Lasius alienus	centipedes such as *Geophilus*, wireworms

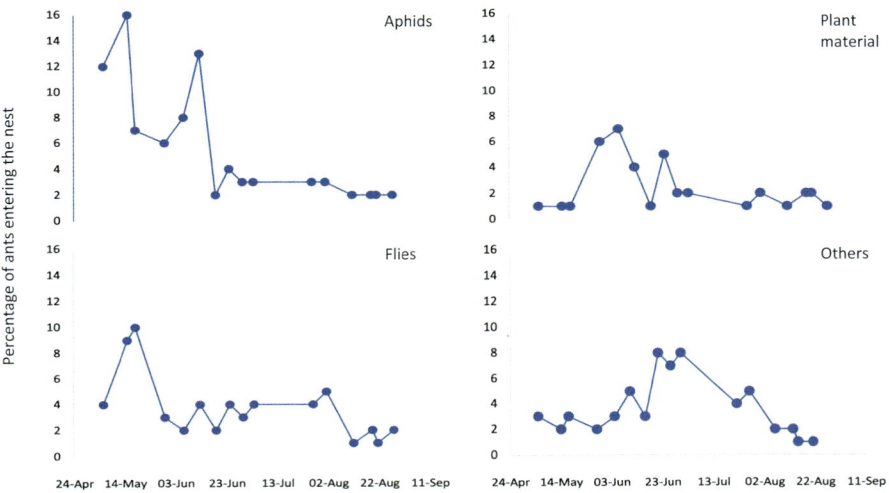

Fig. 2.25 Food income to a nest of *Formica rufa* in Cringlebarrow Woods, North Lancashire over a period of six months in 1973 (from Skinner 1980b)

effect', but this should be minimised wherever possible. One way to reduce disturbance is to install permanent sampling devices on trails (Chapter 7). These do disturb the ants, but they get used to them. Automatic devices make longer-term sampling possible and remove the 'one-observer-one-trail' constraint. The sampling limit depends only on the number of devices available, although sorting the collected material can be very time consuming (Skinner 1980b).

Ants are basically omnivorous animals. All but the most specialised include both animal and vegetable components in their diet. However, ants do not generally eat solid plant material directly but rely on sap-suckers, except for the fascinating case of the fungus growers. The fungus-growing ants collect leaves, grow fungi on them and then eat parts of the fungus. None of these ants occur in Britain or Ireland, although there have been some suggestions that *Lasius fuliginosus* may eat some of the fungus that grows in the carton from which they make their nest (p. 18). The vegetable component of the ant diet consists almost exclusively of honeydew, a sugar-rich fluid excreted by plant-sucking bugs. Ants have sometimes been seen chewing at the bases of bud scales or drinking nectar in flowers, but these habits do not seem to be widespread (Fig. 2.26). Indeed, some flowers appear to have ways of keeping ants away from the nectar; ants are not good pollinators. The situation is different in the extrafloral nectaries on the fronds of bracken fern,

Fig. 2.26 *Formica cunicularia* on an umbellifer, Dungeness, August 2020. This ant can be commonly found on various umbellifers where it hunts but is also known to take nectar.

Fig. 2.27 *Myrmica* species licking nectar from an extrafloral nectary on bracken

Pteridium aquilinum, and at the base of the leaf stalk of broad bean or common vetch. Ants visit these structures to lick the sugary nectar, and the plants evidently derive some benefit because the ants protect them from herbivores (Fig. 2.27). This association is commonly described as a mutualism, but it does not always result in overall benefit for both parties; the balance of costs and benefits for the ant and the plant often depends on local circumstances (Koptur & Lawton 1988, Rashbrook *et al.* 1992). Lawton & Heads (1984) demonstrated that neither *Myrmica* species nor *Formica lemani* showed any protective effect on bracken, but in a later study by Heads (1986), a weak effect of *Formica lugubris* was shown. Further study of the association with this common plant is warranted.

By far the commonest plant-derived food for ants is honeydew from plant-sucking bugs, such as aphids, coccids, membracids, psyllids and phylloxerids. Phloem sap is rich in sugars, particularly sucrose (table sugar), but relatively poor in amino acids. Plant-suckers, notably aphids, require a higher ratio of amino acids to sugars than the phloem provides. They therefore take in much more sugar than they need, and the excess sugar and water is voided from the anus as honeydew (Fig. 2.28). Ants avidly tend aphids to get this

Fig. 2.28 *Lasius fuliginosus* tending the poplar leaf aphid, *Chaitophorus leucomelas*

material. A very wide range of such ant–aphid mutualisms has been observed. Domisch *et al.* (2016) list over 170 tended sap-suckers by members of the wood ant group alone. Some of the major conclusions and unanswered questions of studies of these associations are considered in Chapter 3.

Apart from some very unusual examples, all ants need some animal food in the diet. Again, as with honeydew, species differ in the extent to which they depend on such foods. In the tropics, army ants rely almost entirely on animal food, whereas leaf-cutter ants eat very little, if any. In Britain, all our species are omnivorous to some degree. Little is known about exactly how ants find and catch their prey. It is clear that wood ants can see from a distance of 10 cm or so. Some ants may be able to detect vibrations caused by movement of the prey. All ants have a good sense of smell, which may be important in prey detection. Small invertebrates are caught by a slow approach followed by a final pounce. In those ant species with stings, poison is then injected. In others, the prey is sprayed with poisons from glands. Fellow workers nearby often help to kill the prey. However, many of the prey species seen being carried along ant trails actually have very effective escape mechanisms. This has led to the suggestion that much prey is caught only when incapacitated in some way. Some researchers have suggested that many ant species are acting mainly as scavengers when they bring back such food. This may well be the case for wood ants. There is much scope for further work on the food income of nests, both in terms of what is brought back and in terms of energy flow.

One interesting aspect of food choice, at least in the wood ants, is that they are able to capitalise on a temporary abundance. A dead pheasant was observed to be cleaned of all its flesh over a single season, the ants making a new trail to this food source. In addition, seasonal changes in food income often reflect abundance of that food in the environment. *Formica rufa* was shown to bring back huge numbers of St Mark's flies, *Bibio marcii*, in the spring when they are very abundant in the field (Fig. 2.29).

We know little about the differences in diet between different species and different colonies or changes of diet with season. A summary of some of the conclusions reached for the ant fauna of Britain is given in Table 2.4, above. Much information could be added by careful studies of other species. The data available for *Myrmica* species are very generalised, but other observations led Brian (1977) to expect differences within the genus. For instance, *M. scabrinodis* is

Fig. 2.29 A male *Bibio marcii* (St Mark's fly)

harvester ants
Ants that collect and
store seeds as a primary
food source, such as seed
specialists in the genera
Messor, Pogonomyrmex
and *Pheidole*.

Fig. 2.30 The yellow body
is the elaiosome on a seed
of gorse, *Ulex europaeus*

more of a hunter in short vegetation than its close relative, *M. ruginodis*. Detailed inventories of food have not been made for many species and could be constructed by careful observational work. This requires little equipment, but plenty of patience and time.

In addition to animal prey and honeydew, some ants eat seeds and even collect and store them for later use (e.g. harvester ants). This is a very common phenomenon in ants, especially in desert environments where food, in the form of the seeds of ephemeral desert plants, may be available for only a short time. In Britain, *Myrmica* species are known to collect seeds of tormentil, *Potentilla erecta*, but it is not known whether these are eaten. In the spring *Tetramorium caespitum* collects a wide range of seeds and feeds them to larvae, after chewing them (Brian 1977). In other cases of seed collection noted in the ants of Britain, the seed has an oily body, the elaiosome (or caruncle in the spurges), which the ants consume, leaving the rest of the seed undamaged and thus able to germinate. In these cases, the ants are performing a dispersal function for the plants. For example, seeds of gorse, borage, *Viola* and *Primula* species are known to be treated in this way (Fig. 2.30). In studies of *Formica rufa*, seeds of cleavers, *Galium aparine*, were often seen to be carried by workers (personal observations). These seeds do not have an elaiosome and it is not clear why they are carried, but a possibility is that they are used as nest material (Gorb *et al.* 1997).

NPP

Net primary production is the energy stored in plants from their photosynthetic activity and after the loss due to plant respiration.

Food provides not only materials for growth, development and reproduction, but also energy for all activities. In a study by Fry (1988) at Burnham Beeches, Buckinghamshire, it was shown that the proportion of net primary production (NPP) moved by *F. rufa* workers is between 0.12% and 0.47%. Fry estimated that the other top predator in the woods she surveyed, the fox *Vulpes vulpes*, moved only 0.05% of NPP across its foraging area. She concluded that the energy flows created by *F. rufa* play an important role in the ecology of Burnham Beeches. However, attempts to look at energy budgets of ants' nests are not common. The work of Nielsen (1972) on *Lasius alienus* in Denmark, Golley & Gentry (1964) and MacKay (1985) on *Pogonomyrmex* species, can serve as models and a starting point for anyone interested in doing such study on other species.

2.5 Foraging strategies

Ants use a diverse array of foraging strategies to locate and secure their food. According to Hölldobler & Wilson (1990), the behaviour we call foraging can be split into three main phases or categories: hunting, retrieving and defence. The more primitive ant species tend to have workers that forage alone, carrying or dragging the food to the nest without help. Such workers navigate with the aid of light patterns, gravity and possibly also chemical cues. Wood ants use the disposition and shapes of trees, shrubs and other objects against the sky (Rosengren 1971), together with the sun's direction, chemical cues (Beugnon & Fourcassie 1988) and maybe even magnetic fields (Çamlitepe & Stradling 1995). Some species, notably *Lasius fuliginosus*, rely more heavily on chemical trails (e.g. Carthy 1951).

Many species make long-lasting trails, which they follow from year to year. Memory of foraging sites seems to be good. There is evidence that each worker is responsible for only a restricted part of the territory, so that the area is partitioned amongst the worker population. Thus, each worker has only a small area to remember. There are several ways in which workers recruit help to exploit a newly discovered food source. A worker arriving back at the nest with food from such a source may simply alert others to forage. They then rush out at random, but there is a high chance that some will find the food. A more advanced state is seen when workers follow a successful worker to a new source by 'tandem running'. At an even more advanced level are the likes of *Myrmica* species, *Tetramorium caespitum*, *Lasius niger* and *L. fuliginosus*, which are known to lay scent trails.

Little is known about the trails in these species. For example, are they directional? We know nothing at all about the situation in most of the other species, although there is some suggestion that the parasitic ant *Formicoxenus nitidulus* (which occurs in Britain but not Ireland) follows trails left by one of its wood ant hosts (in this case the species *Formica polyctena*, which is not found in Britain or Ireland) (Elgert & Rosengren 1977). An intriguing question arises as to what *F. nitidulus* does when it is in the nests of non-trail-laying species such as *F. rufa*.

Much useful work could be done on ant trails, in both wild and laboratory colonies, using simple but effective techniques. In interruption experiments, the area in front of the ant is rubbed, for instance with a finger, to interrupt any chemical trail. Ants reaching the rubbed place stop short and there is a delay before they cross it. Similar results can be obtained by rubbing with other implements or by putting Sellotape over part of the trail. In trail displacement experiments, a trail is established over a sheet of paper, and this is removed and replaced, as it is or after cleaning, back to front or with a fresh one. This experimental technique could form the basis of some interesting research, especially on laboratory colonies. Feynman (1985) gives some further details. The trail can be made visible by dusting baby powder or lycopodium powder onto the area where it is thought to have been laid, and then blowing away any powder which is not stuck down. This works well with *L. fuliginosus*. An artificial trail can be made by dragging the abdominal tip of a dead ant along a line to see whether live workers will then follow this trail. If they do, further experiments could explore the efficiency of different glands in producing a followable trail. The glands would need to be separated, by careful dissection, for such experiments.

Some ants will follow trails laid by another species. In interspecific trail-following experiments, different species are tested for their ability to follow trails formed by others. Hölldobler & Wilson (1990) list ants known to be trail formers and followers, together with the source of the trail substance, and provide references. Hangartner (1967) has shown how *Lasius fuliginosus* follows a trail (Fig. 2.31). These ants do not follow the chemical deposited on the ground but walk along the 'tunnel' formed by its vapour, waving their antennae and swaying as they go. A worker moving along the tunnel would naturally tend to stray out of it from side to side as she goes. When the sense organs on her antennae tell her that the antenna has moved out of

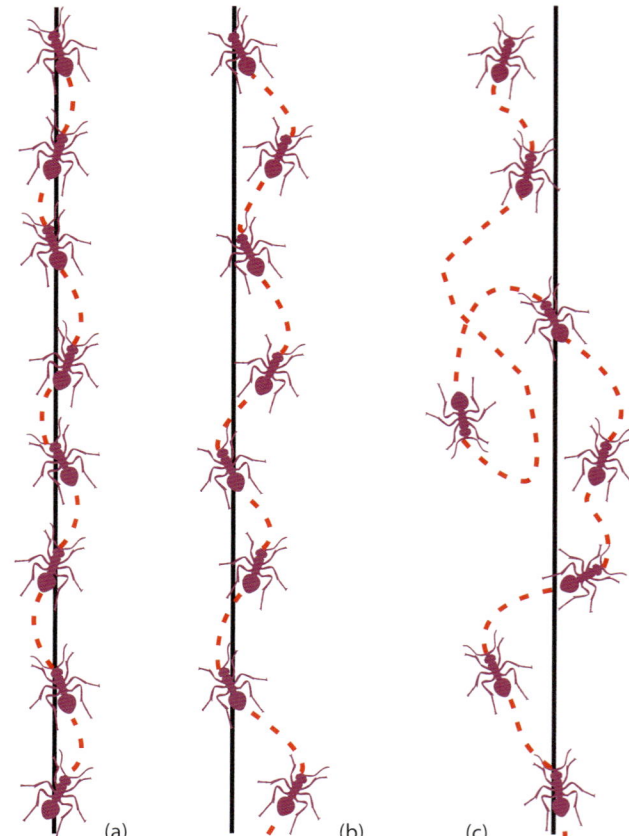

Fig. 2.31 *Lasius fuliginosus* following a trail (see text) with antennae intact (a), or with one antenna missing (b), or with antennae crossed (c) (after Hangartner 1967)

the tunnel, the worker will swing back. Amputation of one antenna leads to overcompensation on the other side. If the antennae are crossed over and fixed in that position, the worker finds great difficulty in orientating, but according to Hangartner she can still do so by a light compass response. All this gives no indication that directional information can be picked up from a trail, so can ants follow a trail in the right direction? It has been suggested that ants lay a trail which is polarised, for example in *M. ruginodis* (MacGregor 1948). In this case, and probably in *L. fuliginosus*, droplets of trail substance appear to be tapered so as to point one way. However, it is not clear how an ant following a 'vapour tunnel' trail, as described by Hangartner, could pick up this information. Some more recent experiments have indicated that directional clues come not from the trail itself, but from other sources. For instance, these clues derive from other workers in the case of *Pheidologeton diversus* (Moffett 1987). There is clearly scope for further contributions in this fascinating area. Hölldobler & Wilson (1990) list a wide range of chemicals used as trail substances. The techniques for isolating and identifying these substances require sophisticated and expensive equipment and are probably beyond the scope of most readers of this book. An excellent summary of this vast subject can be found in chapter 7 of Hölldobler & Wilson (1990).

3 Ants and their environment

biomass
The total weight or mass of organisms in a given area (terrestrial) or volume (aquatic).

With trillions of individuals and a total biomass that equals about 20% of that of all humankind (Schultheiss *et al.* 2022), it is no wonder that ants have a major influence on both the living and non-living environment. In some situations, ants are the major animal component thereof (for example, in tropical rainforest). Chomicki & Renner (2017) provide a useful account of the interactions of ants with animals and plants.

3.1 The biotic environment

Communities of ants

In general, if one species of ant is found in an area, several other species will be found coexisting with it and the species combinations are predictable. Such an assemblage of species is called a community. In a new habitat, some species cannot gain admission because they have not got the dispersal power to reach the habitat. Of those that do colonise, some species will not survive for very long, and others will become established. This process of elimination will produce a relatively stable assemblage, or community.

biotic and abiotic factors
Factors that are living, such as predators, parasites and competitors are biotic, whereas factors that are non-living, such as light intensity, temperature and soil minerals are abiotic.

Survival within the new habitat will depend on both biotic and abiotic factors. Of biotic factors, one of the most thoroughly studied is competition (see, for example, Johansson & Gibb (2016) on wood ants). Savolainen & Vepsäläinen (1988) were the first to propose the now widely studied notion of dominance hierarchies in ants. In temperate forests, they identified dominant species (all the wood ants, together with *Formica exsecta*), encounter species (*Lasius niger*, *Lasius fuliginosus* and *Tetramorium caespitum*) and submissive species (a long list including *Formica fusca* and *Formica lemani*, seven species of *Myrmica* and *Leptothorax acervorum*). The dominant species defend territories, including the resources in them, the encounter species defend just the resources and the submissives just their nest. This linear hierarchy described may, however, be variable depending on such factors as temperature, the size and type of a resource and parasites. Readers are referred to Johansson & Gibb (2016) for more detail and references.

Extensive studies of ant communities were reported by M.V. Brian and his associates at the Furzebrook Research Station in Dorset. The following account, which relies heavily on Brian's work, gives a couple of examples. More details can be found in his original papers (e.g. Brian 1955, 1964) and his books (Brian 1965, 1977 chapter 8, 1983 chapter 16). In a patch of acid grassland, he found that 10% of the total area was covered by nest mounds of *Lasius flavus*, with a nest density as high as 1,500 nests per hectare (Figs 3.1, 3.2). In the most densely populated south-facing regions, the nests were evenly spaced. This apparent mutual avoidance can be taken as evidence for intraspecific competition. Presumably, a queen attempting to found a new nest within the territory of an existing nest would be killed. In some places, particularly on the valley floor, *Myrmica scabrinodis* and *Myrmica rubra* replaced *L. flavus*. Here it was much cooler, and the drainage was not as good as it was on the valley sides because of the nature of the subsoil. Another community was made up of *L. flavus* and *L. niger*. The *L. niger* were found under stones, interspersed amongst the *L. flavus* nests. There was some evidence for direct competition; *L. niger* workers were often encountered with the mandibles of a detached *L. flavus* worker head clamped around one leg. Less direct evidence of interaction between the two species came from the finding that queen production in *L. flavus* was much reduced in

intraspecific competition
Competition for such things as food between members of the same species, whereas interspecific competition is between members of different species.

Fig. 3.1 Nest mounds of *Lasius flavus*

Fig. 3.2 Large nest mound of *Lasius flavus*. Notice how different the vegetation on the mound is from that of its surroundings.

places where *L. niger* was present. On limestone soils in the south of England, Brian found *L. flavus* and *M. rubra* together. Here *M. rubra* could nest only in the shady areas and was absent from sunnier places. The two species were separated in their foraging too. *Myrmica rubra* found food above ground, whereas *L. flavus* fed below ground level.

Dry, base-poor soils in the south of England support a plant community dominated by ling (or heather), *Calluna vulgaris*. A very different ant community exists here. In the best areas, *T. caespitum* (Fig. 3.3) was found with *Lasius alienus* (now known to be *L. psammophilus*, see p. 112) (Fig. 3.4). *Lasius niger* was frequent, and the large black *F. fusca* was present in smaller numbers. In addition, there were occasional nests of *Tapinoma erraticum* (Fig. 3.5), *Formica cunicularia*, *Formica picea* (formerly *Formica candida* and has also been known as *Formica*

Fig. 3.3 *Tetramorium caespitum* workers and brood nesting under a tile trap (Jake Alagoa)

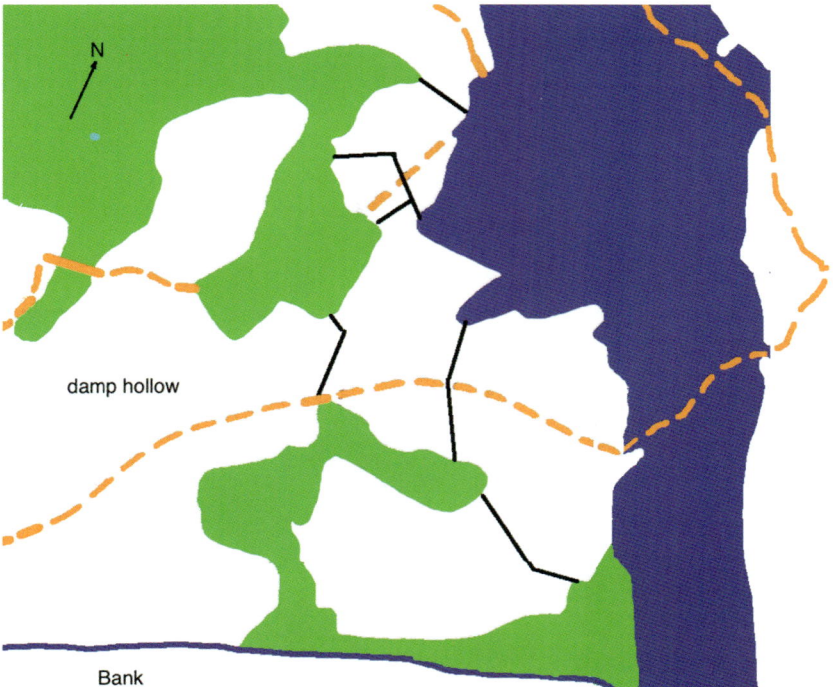

Fig. 3.4 A map of an ant community on a Dorset heath. *Lasius niger* (blue) lives in the shade of a bank and in a damp hollow (orange dashed line), *Lasius alienus* (green) lives on higher ground, and *Tetramorium caespitum* (white) lives in between. The black lines show divisions between colonies within the species. After Brian (1983).

Fig. 3.5 *Tapinoma erraticum* workers

transkaucasica) (Fig. 3.6), *Myrmica ruginodis*, *M. scabrinodis* and *Myrmica sabuleti*. *Lasius flavus* occurred in one or two places. The parasitic ants *Strongylognathus testaceus* and *Tetramorium atratulum* (formerly *Anergates atratulus*) were also occasionally present in nests of the host, *T. caespitum*.

Aside from these, studies on British ant communities are rare. However, Waloff and Blackith (1962) and Elmes (1974) provide some more information on these grassland species. In his 1983 book, Brian deals with communities worldwide and it is clear from this that Britain and Ireland have recognisable communities only in grassland. The steppes and scrub of Eastern Europe and the pastures of the southern USA, on the other hand, do have complex communities and Brian (1983) gives details and references for these studies.

Fig. 3.6 The unique conical nest of *Formica picea* built of small fragments of vegetation around a scaffolding of purple moor-grass

The response of ant communities to disturbance has been widely studied, due to their frequent dominance and ecological importance in most terrestrial ecosystems. In this context they can form very useful bioindicators. This approach has been pioneered mainly in Australia (e.g. Andersen & Majer 2004).

Ants and plants

Ants and plants are associated at many levels in all environments. The most intimate of these relationships are found in the tropics, where some plants produce specialised structures in which ants build their nests, as well as providing the ants with food. In Britain, relationships do occur, but they are much less intimate. In nearly all cases, the association has some mutualistic elements.

Hölldobler & Wilson (1990) note the following strands:

- ants protecting plants
- plants sheltering ants
- plants feeding ants
- ants feeding plants
- ants dispersing plants
- ants pollinating flowers
- ants pruning and weeding
- ants parasitised by plants.

In nearly all situations, ants will forage on plants and collect prey insects, many of which are herbivorous. In the wood ants there is no doubt that this removal of potential herbivores has a protective effect. There was a reduction in the defoliation of trees in the presence of *Formica rufa* when compared with ant-free control trees (Skinner & Whittaker 1981; Whittaker & Warrington 1985). This protection of trees by wood ants has been used as a form of biological control of economically important pest species but is not yet proved to be effective (Robinson *et al.* 2016). The impact of such ant predation on less numerous species of insect herbivore will, of course, be less obvious, but still worthy of investigation.

Worldwide, there are many examples of ants living in foliage or other parts of plants. Some plants form special structures, called domatia. In Britain there are no such examples, but some species do utilise trees in various ways (p. 17).

In the tropics, there is a wide range of plants that produce special food bodies on which ants feed but, again, no such associations seem to occur in Britain or Ireland. However, plants produce food for ants in another way: in the extrafloral nectaries, and these are found in British plant species. The best known are those on bracken, *Pteridium aquilinum* (p. 42). The case of elaiosomes on seeds has already been mentioned (p. 45).

It is well known that more luxuriant plant communities occur close to ants' nests than in the surrounding areas. King (1977a,b,c, 1981) made a careful study of vegetation near nests of *L. flavus*. The favourability of nests as sites for plant growth is probably due in part to the aeration of the soil and enrichment by mineral nutrients from excretion and refuse dumping. Yet this is not the whole story. King showed that some plant species are less frequent on the ant hills than elsewhere. Wood ants, too, have some effect on vegetation near their nests and on soil pH. Further study of the influence of nests on vegetation would be worthwhile.

Although plants in Britain demonstrate few adaptations to housing or feeding ants, many have seeds adapted for dispersal by ants. Some plant species have food bodies on the seed that attract ants, which carry the seeds about and thus disperse them. Seeds of violets, *Viola* species, and gorse, *Ulex* species, are commonly carried about by *Lasius* and *Formica* species (p. 45). *Myrmica* species are known to collect seeds of tormentil, *Potentilla erecta*, and store them in their nest, although we do not know what is subsequently done with them. In colonies of *Tetramorium* species, many seeds are fed to the brood. Beattie (1985) concluded that the main advantage to the plants of this seed transport comes from nutrient enrichment (the seeds are carried to nutrient-rich areas near the nest), and seed predator avoidance, but Hölldobler & Wilson (1990) point out that more data are needed to establish this conclusion with certainty. Observations of ants transporting seeds, and further research on the possible advantages to the seed and seedling, would be very valuable.

prostrate (in a plant)
Lying on the ground.

There is little evidence that ants can act as pollinators (p. 57). It has been suggested that plants with a prostrate growth form and small flowers, together with a number of other features, would seem to be adapted to ant pollination (Hickman 1974), but this idea is now in doubt. There is evidence that these plants are adapted for pollination by very small flying insects rather than by ants. Recent studies have, indeed, shown that ants seem to impair pollen grain function. After contact with ants, some pollen grains fail to germinate and those that do grow produce poor pollen tubes. This may be because ants do not have specialised brood chambers, as bees and wasps do. The suggestion has been put forward that they are therefore vulnerable to attack by fungi and bacteria in the nest. A way of avoiding such attack would be to smear all brood with antibiotics. There is good evidence that ants produce these antibiotics and that they have a negative effect on pollen grains. Thus, ants have evolved to produce antibiotics which protect them against microbial attack and, incidentally, affect pollen. This would exert a strong evolutionary pressure against flowers evolving adaptations for pollination by ants (Beattie 1985; Peakall *et al.* 1991).

However, the jury still seems to be out on the question of whether ants pollinate. In 1998, Puterbaugh published a detailed analysis of the situation in *Formica neorufibarbis gelida*. It has already been pointed out (p. 41 and Fig. 2.26) that *F. cunicularia* in the UK spends a lot of time in the heads of umbelliferous flowers. Since *F. cunicularia* is in the same subgroup of the genus as *F. neorufibarbis gelida*, careful study of the possible pollination of these plants by this ant could be worthwhile. A lookout should be kept for other species of ants in flowers and, where seen, pains taken to see if there is any evidence for pollination.

In the tropics, some ants prune plants overgrowing those on which they are nesting. Janzen (1967) showed that acacia ants do this. Real pruning or weeding has not been demonstrated in British ants, although local effects on vegetation can be seen near the nests of wood ants. The possibility that pruning or weeding is involved is worthy of further exploration. This is especially the case in wood ants, where some preliminary evidence already exists.

Ant guests, predators and parasites

One of the most fascinating aspects of the biology of ants is their association with other animals. There is no other comparable situation in the animal world, except perhaps situations involving human beings. Other social insects – the bees, wasps and termites – do have guests, but not in such variety as the ants. Virtually all these ant guests are arthropods, although members of some other groups are found with ants occasionally. The phenomenon of 'liking ants' (myrmecophily) is very complex and embraces several types. An early classification of these by Wasmann (1894) is still used today, although it has some imperfections and has been abandoned in the recent work of Hölldobler & Kwapich (2022). Wasmann suggested five categories of myrmecophile (Hölldobler & Wilson 1990).

1. Synechthrans (= hostile myrmecophiles (Robinson *et al.* 2016)). The ants treat these in a hostile manner. They are predators of the ants and survive by effectively defending themselves, for instance by running, or by defensive secretions.
2. Synoeketes. These are mainly scavengers which the ants ignore; presumably their neutral odour does not elicit a response from the ants.
3. Symphiles. These, sometimes called 'true guests', are accepted by the ants as if they were part of the colony.
4. Ectoparasites, endoparasites and parasitoids. These are conventional parasites which live on or in their hosts.
5. Trophobionts. These are the insects that provide the ants with honeydew or glandular secretions and are in turn protected and nurtured by the ants.

The first four of these types belong to a larger grouping, called intranidal guests (meaning guests within the nest). In all cases, the ants are passive and are sought out by the guests, and the guests tend to be parasitic or commensal on the host ant. Most trophobionts are extranidal (living outside the nest). They are sought out by the ants, and the associations are of mutual benefit to both partners. The best-known examples of this group are the honeydew-producing insects, which the ants milk. In other cases, the myrmecophile produces secretions that the ants appear to crave. The best known of these are the caterpillars of blue butterflies (p. 65).

British ants have a wide range of guests, and this book cannot deal with them all in detail. Instead, we introduce a few common species of each kind to illustrate the principles and some of the gaps in our knowledge.

The first four of these categories are intranidal guests:

1. Synechthrans (hostile myrmecophiles, hostile persecuted lodgers)

Animals from many groups including arachnids, Acari, Hemiptera, Diptera and other Hymenoptera have been identified as hostile. Robinson *et al.* (2016) list those found with wood ants. The rove beetles (Staphylinidae) are possibly the best known, and most often encountered. Some rove beetles are quite difficult to identify. The book by Harde & Severa (1984) would be a good place to start. Joy (1976) is still very useful, especially if supplemented by Hodge & Jones (1995) and searches on the internet. Finally, Duff (2024) is the most up-to-date guide. Rove beetles often hide in crevices in and near the nest beside trails. They distract attacking ants by appeasement behaviour in which the

commensal
An organism that gains benefit from a host without doing any harm to it.

Hemiptera
An order of insects, the true bugs, that includes the cicadas, aphids, planthoppers and leafhoppers, assassin bugs, bed bugs and shield bugs.

Diptera
An order of insects, the two-winged flies.

Hymenoptera
An order of insects that includes the ants, bees, wasps and sawflies.

rove beetle
A family of beetles most clearly distinguished by their short wing cases (elytra).

beetle thrusts its abdominal tip toward the attacking ant, and the ant licks it, usually losing interest in its attack. If this fails, the beetle can secrete a further droplet from the abdomen tip which the ant then licks. When the ants are less aggressive the beetle may escape by feigning death. When it is in this state the ants may pick it up, but they usually discard it later, unharmed. If all else fails, the beetle responds by curling the abdomen forward and thrusting it into the ant's face, at the same time releasing a repellent secretion. In addition, some of these beetles have a very ant-like appearance, which may help to protect them from the ants or from other enemies (Donisthorpe 1927b). One abundant British species of this kind is a rove beetle called *Pella humeralis*. Donisthorpe (1927b) referred to it by the older name of *Myrmedonia humeralis* and described it as attacking and killing *F. rufa*, but Hölldobler & Wilson (1990) report research which suggests that these beetles are largely scavengers. That work would bear repetition in the UK. This beetle is also found with *L. fuliginosus*. In a study of *F. rufa* and *P. humeralis*, pitfall traps (p. 279) were set out at different distances from a nest. This beetle was by far the most common animal caught after the ants themselves (Fig. 3.7). In addition, its distribution was correlated with

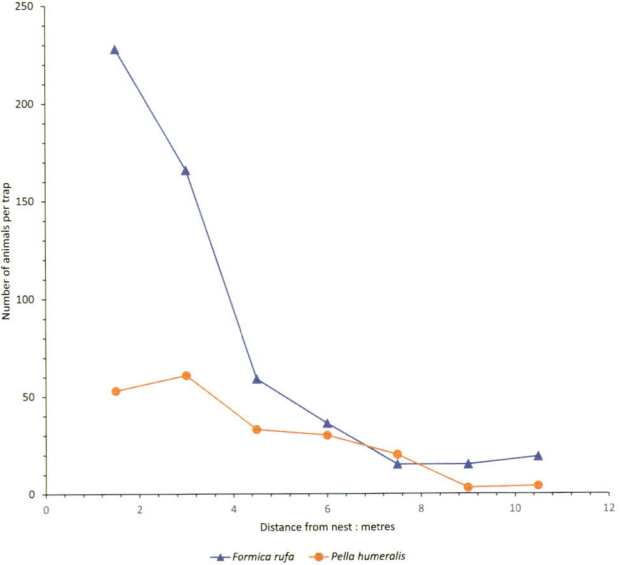

Fig. 3.7 The numbers of *Pella humeralis* and *Formica rufa* caught in pitfall traps around one nest of *F. rufa* over one day

that of the ants. Many other hostile myrmecophiles exist and a careful observer could learn much about their biology. Hölldobler & Kwapich (2022) give a very detailed account of the relationship between various *Pella* species and ants.

2. Synoeketes (indifferently tolerated lodgers)

A more advanced condition than that of *P. humeralis* is found in *Dinarda* species (Donisthorpe 1927b), another sort of staphylinid beetle recorded with *Formica* species in Britain. The larvae of these beetles feed on dead ants and debris and appease the ants in the way described for *P. humeralis*. The adults roam around within the nest stealing prey brought back by the ants and, perhaps, feeding on mites and the ants' eggs and larvae. They are known to steal food being passed from one worker to another and even to beg food from individual workers. As a note of caution, it should be pointed out that Robinson *et al.* (2016) classify the two *Dinarda* species that occur with wood ants as hostile (synechthrans). Donisthorpe gives accounts of a few other beetles in this category; treatment of all of these is beyond the scope of this book. However, it is worth pointing out that one ladybird, *Coccinella magnifica* (*Coccinella distincta* in Donisthorpe) is associated with wood ants. Donisthorpe mentions this species, although gives no details of any benefits that either the ant or the beetle gain from the relationship. It is clear that the ants recognise this beetle as different from the very similar 7-spot ladybird (*Coccinella septempunctata*) as the former is not attacked when placed on a nest, the latter very vigorously so.

Another species of special note is the beetle *Protaetia metallica* (*Cetonia cuprea* in Donisthorpe and *Protaetia metallica* subspecies *cuprea* in some sources) (Fig. 3.8). The larvae of this metallic chafer live with wood ants, apparently feeding off nest material.

A synoekete likely to be very frequently found is the white, blind woodlouse *Platyarthrus hoffmannseggii* (Fig. 3.9). This common species has been little studied, although there is an excellent account of some aspects of its ecology by Brooks (1942) and a provisional report of work by Williams & Franks (1985). Some believe that *P. hoffmannseggii* will live with all ants in its range (so-called pan-myrmecophily) (e.g. Wasmann quoted in Brooks 1942). However, Brooks (1942) and, more recently, Hames (1987) provide data to suggest that this is not the case, and that there is some selection. This would warrant further study, as would other aspects of the ecology of this unique myrmecophile. Indeed,

Fig. 3.8 *Formica aquilonia* workers carrying an adult of the chafer beetle *Protaetia metallica*

Fig. 3.9 The myrmecophilous woodlouse *Platyarthrus hoffmannseggii* in a nest of *Lasius flavus*

it is still not even clear how *P. hoffmannseggii* feeds, with ant faeces, spores of 'lower plants' and general detritus suggested. They may 'milk' aphids and feed on pellets from the infrabuccal pouches of ants. These pouches filter solid particles out of the food the ants imbibe. This material is then disgorged as a pellet, and it is upon these that the *P. hoffmannseggii* may, in part, feed.

Other synoeketes encountered are from the Hymenoptera, especially proctotrupid wasps (which are very ant-like), some Diptera such as Syrphidae (hoverflies) and Phoridae (scuttle flies), Heteroptera (typical bugs), again with some ant-mimicking forms (e.g. *Myrmecoris gracilis*). In addition, Donisthorpe lists springtails, millipedes, pseudoscorpions, spiders (including the fascinating *Theridion* species, which sit on twigs above ant trails and haul ants from below on a silken thread), nematode worms and even the slow worm, *Anguis fragilis*, as being myrmecophiles. This is a vast subject and was very well reviewed by Donisthorpe (1927b). However, nearly a century has passed and, apart from a brief chapter in Brian (1977), not much had advanced, outside the very specialist literature, until publication of the work by Hölldobler & Kwapich (2022).

3. Symphiles (true guests)

According to Donisthorpe (1927b), only five beetles, all staphylinids, fall into this category (Table 3.1). The clearest feature that distinguishes a species as a true guest is that it is fed by the ants with which it lives. This is true of all those in Table 3.1. The commonest and most widespread of these beetle true guests is probably *Lomechusa emarginata*, found

Table 3.1. True guests of ants in Britain (based on Donisthorpe 1927b)

Guest	Recorded hosts	No. of National Biodiversity Network (NBN) Atlas records/ year of most recent record
Claviger testaceus	*Lasius flavus, L. niger, L. alienus, Myrmica scabrinodis*	27/2006
Claviger longicornis	*Lasius niger, L. mixtus, L. umbratus*	10/1921
Lomechusa emarginata	*Formica fusca* (summer), various *Myrmica* species in winter	79/2022
Lomechusa paradoxa	*F. fusca* (summer), various *Myrmica* species in winter	2/2001
Lomechusoides strumosus	*Formica sanguinea*	5/1975
Amphotis marginata	*Lasius fuliginosus*	9/2020

* Donisthorpe was equivocal about the status of *A. marginata*; however, Hölldobler & Kwapich (2017) provide evidence that its behaviour comes under the definition of a symphile.

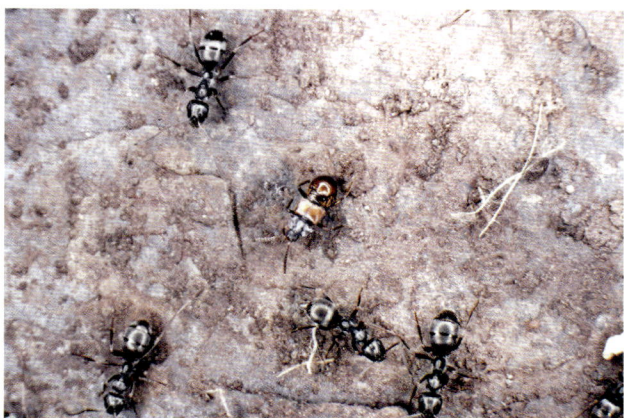

Fig. 3.10 *Lomechusa emarginata* (formerly *Atemeles emarginatus*) with *Formica fusca*

with *F. fusca* in summer and *Myrmica* species in winter (Fig. 3.10). All five of these symphile species are restricted in range, rare in the north of England and apparently absent from Scotland and Ireland. It would be useful to look out for any of them when the nests of the relevant ant species are being observed.

4. Ectoparasites, endoparasites and parasitoids

Not much is known about the parasites of ants, although a number of fungal parasites have been identified around the world, most famously the pan-tropical *Ophiocordyceps unilateralis*, the so-called zombie-ant fungus (e.g. Evans *et al.* 2011).

summit disease
A variety of fungal pathogens cause infected insects to climb to the top of plants, hold on by their jaws, and eventually die, at which point, fungal spores are released.

Boer (2008) reported an instance of summit disease in *F. rufa* in the Netherlands. Workers, found with their jaws clamped on plant stems (usually *Carex* species), were shown to be infected with the fungus *Pandora myrmecophaga*. Field and laboratory observations showed that an infected worker climbs a plant stem '(like a) drunken person'. Eventually, she clamps her jaws around the top of the leaf and dies a few hours later. Fungal rhizoids further glue the ant to the stem and, in a day or so, the fungus grows through thinner parts of the cuticle between segments. From here, wind disseminates the fungal conidia (spores).

In another case in wood ants, a fluke called *Dicrocoelium dendriticum* causes a very similar behaviour. In this instance, the definitive host – cattle or sheep – defaecates and the faeces, containing fluke larvae, are eaten by various snails. The snails wall off the larvae and excrete them as cysts. These get eaten by ants whose behaviour becomes modified.

Infected workers climb to the top of plant stems at night and stay until dawn. If the grass, along with the ant, has not been eaten by a grazer, the ant will return to its normal life in the daytime. This behaviour continues each night until the grass, along with the ant, is eaten by a sheep or cow, at which point the fluke is back in its definitive host.

In 2019, Bos *et al.* reported on their fascinating laboratory experiments on 12 species of European ants, eight of which are also UK species. They subjected the ants to the fungal species *Beauveria bassiana* and *Metarhizium brunneum*. The effects varied according to species. For example, *Myrmica schencki* and *Lasius platythorax* were relatively unaffected, whereas *Formica exsecta* and *L. flavus* proved to be very susceptible. In all cases, however, they concluded that generalised pathogens such as *B. bassiana* and *M. brunneum* are not major sources of mortality in ants. In their study, apart from the effect on ant mortality, they also looked at the measures the ants took to protect themselves. Ants have metapleural gland secretions, a tough cuticle and physiological and/or behavioural responses to protect themselves. Ant societies, like all societies, have increased susceptibility to disease transmission (the COVID-19 pandemic that started in 2020 is an example of rapid disease spread in human society), so

metapleural gland
A gland at the front of the mesosoma that is unique to ants. It secretes antibiotics which are essential in social insects where contact with microbes is likely.

Fig. 3.11 A *Formica aquilonia* worker carrying ectoparasitic mites

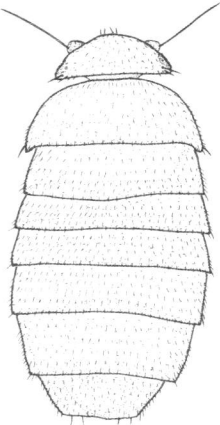

Fig. 3.12 *Aenigmatias lubbocki*, a female scuttle fly (1.6 mm in length), with legs omitted (from Disney 1983)

it is not surprising that ants have evolved a sophisticated range of immune responses. Bos *et al.* (2019) listed these and investigated how exposure to pathogens influences some of them. Their methods are elegant and simple and could well be adapted and used on further species in the UK, with exposure to other pathogens.

Several animals living with ants are ectoparasitic. Franks *et al.* (1990) report on the obligate ectoparasitic mite *Antennophorus grandis*. This mite rides on the ant's body and either takes food that is being passed from worker to worker or solicits it from its host worker. These authors showed that the mites greatly restrict the behaviour of the ants that bear them. Most of the observations were made on colonies kept in formicaria. Fig. 3.11 shows a particularly heavily infested worker. Donisthorpe mentions 39 species of mites associated with ants. Work on these species, along the lines of the study by Franks *et al.* (1990), could be very worthwhile.

A group of insects, the endoparasitoids, specialise in laying eggs inside the bodies of others, which their larvae then consume from within, killing the host in the process (Fig. 3.12). Much remains to be found out about the endoparasitoids. Table 3.2 summarises some of the knowledge we have so far.

There are also the extranidal guests. Associations between ants and sap-sucking insects are described in more detail later (p. 71). Here, we consider the very interesting relationship between ants and the larvae of blue butterflies (family Lycaenidae) (Table 3.3).

It is the large blue's association with ants in England that has been most widely studied, not least because the butterfly recently declined and became extinct in this country and a detailed study of its biology was needed as a basis for plans to reintroduce it. The butterfly larvae feed on the flowers of wild thyme, where they are attended by ants. When a larva reaches the fourth instar, it leaves its food plant and

Table 3.2 Some of the known parasitoids of ants in Britain

Parasitoid	Ant hosts	Notes and references
Hybrizon buccatus	Unknown species	Little is known
Neoneurus species	Unknown species	Little is known, Shaw & Huddleston (1991)
Aenigmatias lubbocki	*Formica picea, F. sanguinea, F. fusca*	Parasites of ant pupae, Disney (1994)
Pseudacteon formicarum	*Myrmica lobicornis, Lasius* species, *F. sanguinea*	Parasites of adult ants, Disney (1994)
Pseudacteon brevicauda	*Myrmica* species	Disney (1994)

Table 3.3 Lycaenid butterflies that have been found in ants' nests (more details are in Thomas & Lewington 1991). Further examples may be revealed by careful study of other lycaenid species.

Common name	Scientific name
Green hairstreak	*Callophrys rubi*
Purple hairstreak	*Neozephyrus quercus*
Silver-studded blue	*Plebejus argus*
Common blue	*Polyommatus icarus*
Chalkhill blue	*Lysandra coridon*
Adonis blue	*Lysandra bellargus*
Large blue	*Phengaris arion*

hides until it is found by a worker of the host ant. The ant then 'milks' the caterpillar, which changes shape, retracting the head and expanding the thoracic segments to assume a humpbacked appearance (Fig. 3.13). Perhaps mistaking it for an ant larva, the ant picks it up and takes it back to the nest. Here, the butterfly caterpillar feeds voraciously on ant larvae, then pupates in the nest and emerges as an adult butterfly the next year. The natural history of this association is beautifully described by Thomas & Lewington (1991), and how this study helped in the conservation and successful reintroduction of the butterfly is described in Thomas *et al.* (2009).

Fig. 3.13 Larva of the large blue butterfly, *Phengaris* (formerly *Maculinea*) *arion* being adopted by *Myrmica sabuleti* (Jeremy Thomas)

Predators

Some of the myrmecophiles dealt with above, such as the caterpillars of certain Lycaenidae, are clearly predators of the ants or their brood. This section deals with some others. Amongst the arthropods are a group of spiders in the family Theridiidae that sit on branches above trails, dropping on workers and wrapping them in silk. Ants are no doubt eaten by many generalist insect predators too, and much remains to be found out about the quantitative aspect of such predation. Workers are eaten by a wide range of bird species, the best known being the green woodpecker (*Picus viridis*). This bird can often be seen probing around in nests of many species during the winter (see p. 22 for more details).

Social parasitism

Several species of ants depend on other ant species to a greater or lesser degree. The importance of temporary social parasitism in nest-founding has been discussed earlier (p. 31). The range of other relationships is illustrated in Table 3.4.

Table 3.4 Species relationships and social parasitism in ants (after Hölldobler & Wilson 1990).

Type	Details	Examples in Britain and Ireland
A. Compound nests Species nest together, but keep brood separate		
Plesiobiosis	Species nest close together, but with little communication	Many ants could be regarded as loosely associated. *Lasius flavus* and *Formica fusca* have recently been shown to commonly associate in this way.
Cleptobiosis	Robber ants: small ants nest beside larger species and feed on refuse or rob returning workers	Probably none
Lestobiosis	Thief ants: small ants share nests with larger species and steal food or prey on brood	*Solenopsis fugax*
Parabiosis	Two species nest together, and may share trails, but keep brood separate	None
Xenobiosis	So-called guest ants inhabit nests of another species and solicit food but keep brood separate. These are probably parasites (although this view is controversial)	*Formicoxenus nitidulus*
B. Mixed colonies Brood mingle and are cared for communally		
Temporary social parasitism	A fertilised queen of one species enters the colony of another and assassinates the host queen, founding her own colony	Several *Lasius* and *Formica* species (see p. 31)
Dulosis (slavery)	One species captures brood of another to rear as slaves	*Formica sanguinea* (see p. 69)
Inquilinism (permanent parasitism)	Parasite spends entire life in host nest; often workerless	*Myrmica hirsuta*, *Myrmica karavajevi*, *Tetramorium atratulum*, *Strongylognathus testaceus*

Fig. 3.14 *Formicoxenus nitidulus* worker (right) with *Formica lugubris* worker (left)

Guest ants

The only guest ant in Britain is *Formicoxenus nitidulus*, which is found with *Formica rufa*, *F. aquilonia* and *F. lugubris* (Fig. 3.14). This species has not been found in Ireland, but host species do occur (*F. lugubris* and *F. aquilonia*), so it might be worth looking in their nests (Niechoj 2011). The colonies are small, and the castes are not well differentiated. The ants remain in the nest of the host nearly all the time. They appear to feed by soliciting from the host or by stealing food from two mutually feeding host workers. The wood ants ignore the guests most of the time, but occasionally attack. The response of the guest is to remain motionless, although in rare cases guests have been observed to sting the host. The relationship is almost a parasitic one, and indeed Hölldobler & Wilson (1990) consider it to be so, but Dumpert (1978) does not agree.

Thief ants

The only thief ant occurring in Britain is *Solenopsis fugax*. Its workers excavate narrow holes in the nest of the host. They then seize host brood, take them into the small holes where they cannot be followed, and eat them. They live with a wide range of species of *Formica* and *Lasius* in Britain. This species is not found in Ireland.

Slavery

The large wood ant-like *Formica sanguinea* (Fig. 3.15) is our only slave-maker ant. It is thought to be a facultative slave-maker, as colonies are commonly found without slaves. *Formica fusca*, *F. lemani*, *F. cunicularia* and *F. rufibarbis* are all potentially enslaved by *F. sanguinea*. Wheeler (1910) described a *F. sanguinea* raid as follows:

'*The sorties occur in July and August after the marriage flight of the slave species has been celebrated and when only workers and mother queens are left in the formicaries. According to Forel the expeditions are infrequent, "scarcely more than two or three a year to a colony". The army of workers usually starts out in the morning and returns in the afternoon, but this depends on the distance of the* sanguinea *nest from the nest to be plundered. Sometimes the slavemakers postpone their sorties till three or four o'clock in the afternoon. On rare occasions they may pillage two different colonies before going home. The* sanguinea *army leaves its nest in a straggling, open phalanx sometimes a few metres broad and often in several companies or detachments. These move to the nest to be pillaged over the directest route over the often numerous*

Fig. 3.15 *Formica sanguinea* worker showing the notched clypeus, which is arrowed (see Key A couplet 10, p. 236)

obstacles in their path. As the forefront of the army is not headed by one or a few workers that might serve as guides, it is not easy to understand how the whole body is able to go so directly to the nest of the slave species, especially when this nest is situated, as is often the case, at a distance of 50 or 100 m ... When the first workers arrive at the nest to be pillaged, they do not enter it at once, but surround it and wait till the other detachments arrive. In the meantime the fusca *or* rufibarbis *scent their approaching foes and either prepare to defend their nest or seize their young and try to break through the cordon of* sanguinea *and escape. They scramble up the grass-blades with their larvae and pupae in their jaws or make off over the ground. The sanguinary ants, however, intercept them, snatch away their charges and begin to pour into the entrances of the nest. Soon they issue forth one by one with the remaining larvae and pupae and start for home. They turn and kill workers of the slave species only when these offer hostile resistance. The troop of cocoon-laden* sanguinea *straggle back to their nest, while the bereft ants slowly enter their pillaged formicary and take up the nurture of the few remaining young or await the appearance of future broods.'*

Although the details of the raid are well known, colony founding in *F. sanguinea* has never been observed in nature.

Fig. 3.16 The degenerate slave-maker *Strongylognathus testaceus* (Phillipe Hoenle)

Laboratory observations, though, suggest that this is achieved by a form of temporary social parasitism, similar to that already described for *F. rufa* (p. 31) (Mori & Le Moli 1998).

Strongylognathus testaceus belongs to a genus of slave-making species, but this species appears to have lost the ability (it is known as a degenerate slave-maker) (Fig. 3.16). The queens of host (*Tetramorium caespitum*) and parasite live side by side. The host queen supplies workers, and large colonies are generated. The parasite produces queens and workers, but these have never been observed to go on raids. *S. testaceus* is therefore considered to be a host-queen tolerant inquiline. This condition is somewhat intermediate between slave-makers and inquilines (Guillem *et al.* 2014). If colonies can be located, some very interesting and original observations could be made. Bolton & Collingwood (1975) believe that *S. testaceus* has a wider range than is indicated by the few existing records from the New Forest, Dorset and Devon. Sites where its host, *Tetramorium caespitum*, is very common (see Map 61 on p. 203) would be well worth investigating for this distinctive species.

Inquilinism

inquiline (in ants)
An obligate social parasite.

The inquiline ant species found are listed in Table 3.4. Among a wide range of adaptations for inquilinism listed by Hölldobler & Wilson (1990), the most obvious is the loss of workers. All three species are workerless. *Tetramorium atratulum* lives with *Tetramorium caespitum*. The strange males never develop fully and are described as pupoid for this reason. As far as we know, they fertilise the females in the natal nest. These females then fly off to find a new host colony whose own queen has died, and walk into it, usually unopposed. There the parasite queen lays many thousands of eggs. With no host queen present to produce workers, however, the colony eventually dies out. *Myrmica karavajevi* is workerless too. It allows its host (in Britain *Myrmica scabrinodis* and *M. sabuleti*) to continue to produce workers, but no sexual forms. The sexuals of *M. karavajevi* are winged and mate outside the nest.

Myrmica hirsuta is found with *Myrmica sabuleti* in Britain. It is known to parasitise *Myrmica lonae* in Finland, so might be worth looking for at the *M. lonae* sites now known from Scotland (p. 155). Neither species is found in Ireland.

Mutualism with aphids and other 'plant lice'

All aphids produce honeydew, but not all species of aphids are tended by ants. This was shown, for example, in the

Fig. 3.17 The tended aphid *Periphyllus testudinaceus*

Fig. 3.18 The non-tended aphid *Drepanosiphum platanoidis* (R. Dransfield & R. Brightwell, influentialpoints.com)

wood ant *Formica rufa* and some aphids on sycamore trees in a northern English woodland. The two aphid species frequent on sycamore leaves were the common periphyllus aphid, *Periphyllus testudinaceus*, which ants tended, and the sycamore aphid *Drepanosiphum platanoidis* (sometimes called *D. platanoides*), which was not tended (Figs 3.17, 3.18). Tending affects aphid numbers. Fig. 3.19 shows how the numbers of the two species changed over a season when ants were kept away from them by sticky bands around the stems. In this simple experiment, the tended species benefited enormously from the presence of the ants, whereas the non-tended species did not. Indeed, the non-tended species suffered from the ant attention. The reason for this was suggested by looking at the prey items that the ants were carrying back along trails to the nest. Many specimens of the sycamore aphid were taken as prey. As indicated, this kind of exclusion experiment is quite simplistic. Styrsky & Eubanks (2007) have produced an excellent review of the literature in this field showing, among other things, the limitations of this simple exclusion approach.

Many questions remain. Why do ants attend one species and not another? Is it because *P. testudinaceus* lives in colonies, so that foraging ants can 'milk' honeydew from several at a time, whereas individuals of *D. platanoidis* are more widely scattered? And how does ant attendance enable aphids to do better than they would if the ants were not there? If honeydew were not removed by the ants, it would accumulate around the aphids, causing physical problems and enhancing the growth of mould, which could be detrimental to the aphids and the plants that support them. In other species, which are sometimes tended and sometimes not, droplets of honeydew are flicked away from the body with the hind legs or the tip of the abdomen, or by contraction of the abdomen or rectum. Some ants, such as members of the genus *Leptothorax*, will sometimes lick up such honeydew from the ground or from leaf surfaces. Others, such as the tropical genus *Acropyga*, are totally dependent on honeydew (Sudd 1967). *Lasius* species are probably the group most dependent on honeydew in Britain and Ireland, although species of *Formica* and *Myrmica* take a considerable amount. Some aphids, such as *Forda formicaria*, are found only in association with ants, and others do not thrive in their absence. For instance, the sycamore-dwelling aphid *P. testudinaceus* is never very common except when attended by ants.

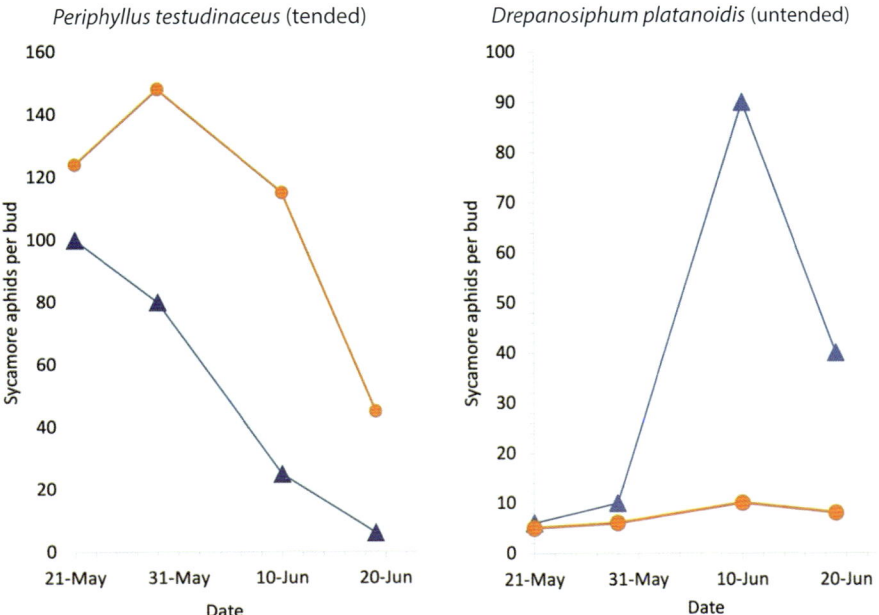

Fig. 3.19 Effects of ants on a tended aphid (*Periphyllus testudinaceus*) and an untended aphid (*Drepanosiphum platanoidis*) on sycamore. Ants had free access to some trees (orange) and were excluded from others by sticky bands (blue). From Skinner & Whittaker 1981.

Attended aphids have adaptations for ant attendance, although Stadler & Dixon (2008) state that some ant-attended aphids appear to have undergone little morphological change compared with other ant-attended species. One of the most obvious adaptations to ant attendance involves the cornicles or siphunculi. These are paired structures on the aphid's abdomen, used in defence (Fig. 3.20). Compare the long cornicles of *U. cirsii* (Fig. 3.20) with those of *P. testudinaceus* (Fig. 3.17).

When some aphids are approached by a potential predator, a waxy substance is exuded from the cornicles, and this usually deters further attack (Dixon & Theime 2020). In many (but not all) frequently tended species these cornicles are very small. Some aphids will also respond to potential predators by kicking, walking away or even dropping from the plant (Rotheray 1989, 1994). In species

Fig. 3.20 *Uroleucon cirsii* on thistle showing the cornicles

that allow ant attendance, these behaviours seem to be suppressed. A further adaptation involves the so-called trophobiotic organ, a basket of bristles around the aphid's anus which serves to hold the honeydew droplet for the ants.

As we have seen, ant attendance increases population numbers of aphids such as *P. testudinaceus*. In colonies of aphids of the genus *Aphis*, ant attendance reduces the tendency to produce winged forms (alates). This, in turn, allows the colony to grow larger by preventing dispersal. Some researchers believe that this effect may be due to the presence, in the secretions of the ants, of juvenile hormone (JH) mimics. JH promotes winglessness, and similar molecules in ant secretions may have the same effect (Kleinjan & Mittler 1975). This has been noted in only a restricted range of ant–aphid associations. In other situations, ants may nibble the wings of alates, preventing them from flying away (Kunkel 1973).

The most obvious way in which ants increase aphid populations under their care is by removing, or deterring attack by, the aphids' natural enemies. This effect can usually be shown very clearly by simply comparing the numbers of such enemies in tended and non-tended situations (although see the paper by Styrsky & Eubanks 2007). Although it is natural to suppose that the aggressive behaviour of the ants leads to a reduction in the numbers of aphid predators and parasites, not much quantitative work has been done on this aspect of ant–aphid mutualisms. Here is a rich field for research, both in the wild and in laboratory colonies of aphids on potted plants. A productive approach to such studies is illustrated by some of the work that has been done by, for example, Way (1954) and Bristow (1984). Way (1954) used several experimental techniques. The attendant ant, in this case *Oecophylla longinoda*, was excluded from colonies of plant-sucking bugs, in this case a scale insect *Saissetia zanzibarensis*, either by completely removing the ants or by ringing the plant stem with bands of sticky grease which the ants could not pass. Some of Way's experiments involved potted plants supporting the scale insect and the ant. As mentioned, Styrsky & Eubanks (2007) give an excellent review of this field.

Another way in which ants appear to aid aphids is by removing the potentially harmful honeydew. In the association that he studied, Way showed that the scale insect did much better when attended by ants, but that the benefits of attendance could be partially mimicked by washing away the accumulated honeydew with water.

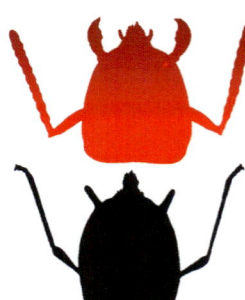

Fig. 3.21 Upper: the head of an ant; lower: the rear of an aphid (after Kloft 1959)

Homoptera
A suborder of the order Hemiptera. The term has been superseded for classification because it is no longer considered to represent a single evolutionary group, but it is used throughout this book as a convenient term for aphids, scale insects, cicadas and leafhoppers, all with sucking mouthparts and often tended by ants for the sugary honeydew they produce.

A further component of the ant effect involves the rate of reproduction of the aphids, which in the tended species seems to be increased in the presence of ants (e.g. El-Ziady & Kennedy 1956). Douglas & Sudd (1980) give a very detailed account of the relationship between the ant *Formica lugubris* and the aphid *Symydobius oblongus* on birch. Careful observations of ant and aphid showed that although the aphid was tended for only 14% of the time, it released 84% of its honeydew during this period. A touch of the aphid's abdomen by the ant led almost invariably to the production of a droplet; the idea that ants 'milk' aphids was thus appropriate in this case. This study also showed that the aphid could communicate with the ant, signalling its readiness to release a droplet by raising the abdomen. In its partnership with the ant *Formica polyctena*, the aphid *Lachnus roboris* conveys the same message by raising its back legs. Similar careful studies on other ant–aphid partnerships are likely to yield useful information.

Ants are normally very aggressive, and it is still not clearly understood how this behaviour is suppressed in their relationship with aphids. Kloft (1959) suggested that the rear of an aphid's abdomen resembles an ant worker offering food (Fig. 3.21). The ant might mistake the combination of features on the aphid's rear for the front end of a sister worker and switch into a behaviour pattern resembling the normal food-sharing that occurs between sister workers. Kloft suggested that the normal leg waving or kicking that most aphids perform when disturbed has become ritualised as an interspecific signal. On the other hand, many non-aphid Homoptera, such as scale insects and mealy bugs, which also produce honeydew, are totally unlike ants in appearance, but are just as avidly attended. Perhaps the resemblance noted by Kloft is simply coincidental. Stadler & Dixon (2008) suggest that this theory lacks both empirical and theoretical evidence. The whole idea needs to be tested by careful observation and experiment.

A further way in which aggressiveness may be suppressed is by chemical communication. Some very closely ant-associated aphids are known to produce substances other than honeydew, which are attractive to ants (Hölldobler & Wilson 1990). It is also possible that ants scent-mark aphid colonies. Both these ideas need further investigation.

Stadler & Dixon (2008) provide a relatively up-to-date and comprehensive review of the subject of mutualism between ants and their insect partners.

3.2 The abiotic environment

Due largely to their sheer numbers, ants can be thought of as important ecosystem engineers. One aspect of this is their effect on the soil and nutrient cycles.

ecosystem engineer
A species that modifies its environment, also can be thought of as a keystone species.

keystone species
A species that is important in defining what the ecosystem is like, without it the system would change dramatically.

Ants and the soil

Many ants live in the soil and build their nests there (pp. 272–276). It seems likely, then, that they might have a significant influence on soil structure, just as earthworms do, particularly because of the sheer numbers of ants in certain habitats and their habit of moving organic material about. The effects of ants on soils are both physical and chemical.

Physical effects

Studies in which active and inactive nest sites are compared with undisturbed soil nearby have concentrated on the effects of mound-building ant species. They show that ants transfer soil particles differentially, with the clays and silts being most commonly moved (Wiken *et al.* 1976). Baxter & Hole (1966) showed that ants (*Formica cinerea* in the USA) moved 7.4 tonnes per hectare of soil per year in mound-building operations. Is this of any great significance? Clearly, in the short term the movements are very localised near the nest, but many nests are short-lived and so in the longer term much of the habitat may be affected. Baxter & Hole estimated that *F. cinerea* could move a 15 cm layer of subsoil to the surface in a period of 106 years, on the reasonable assumption that they stayed in the same nest for about 10 years. In a related study, Salem & Hole (1968) credited ants with the conversion of a forest soil to a prairie soil. In Europe, Czerwinski *et al.* (1971) suggested that *Myrmica* species move to a fresh nest site twice, or occasionally more often, in a single season and thus spread their effects on soil further than one might at first suppose. A note of caution comes from another USA study in which Wiken *et al.* (1976) demonstrated that soil in unoccupied mounds was very similar to that in undisturbed areas, and so concluded that the ant effect is only short-lived.

Chemical effects

Ants are associated with the production of formic (methanoic) acid as a defence or alarm substance. So, do ants affect soil pH? Research has focused on the mound-building ant *L. flavus*. Most of the studies indicate no significant difference in pH between mounds of this species and the surrounding soil, except in acidic grassland where ant activity seems to increase pH. There is no evidence at all for the expected decrease in pH. The increase of pH in acidic grassland is probably caused by accumulation of basic material in nests.

It has also been suggested that ants affect the nutrient status of the soil, and here there is more evidence. The most frequently reported effects involve phosphorus, potassium and carbon. In nests of *Myrmica* species, these elements have been shown to be increased by as much as four times. Simple chemical soil analysis kits are easily obtainable and much useful work could be done with these in and around nests.

Little work has been done on British or Irish ants, aside from that mentioned above by King. This is now a huge subject and interested readers are referred to Farji-Brener & Werenkraut (2017), where many further references will also be found. Frouz & Jilková (2008) provide a more general review.

4 Distribution and abundance

Ants are found throughout the world, with nearly 16,000 species named so far (Schultheiss *et al.* 2022). Most of these are tropical or subtropical. In Europe there are about 400 species (Lebas *et al.* 2019). In Britain, Ireland and associated islands (the region covered by this book, see p. 3), there are just around 60 outdoor-living species recognised (native or well-established introductions), of which six or seven are found only in the very southerly Channel Islands. This small list includes many species that are at the northern edge of their range. For instance, the distribution of the very southern (in British terms) *Tapinoma erraticum* (Fig. 4.1) is typical of about 15 species on the British and Irish list used in this book. Therefore, there are more species as one moves to the south of Britain. This is illustrated by the numbers of species featured in local guides. Macdonald (2013) lists 21 species for the Scottish Highlands, Collingwood & Hughes (1987) list 17 species in Yorkshire, and Hargreaves & White (2021) lists 18 species in Lancashire. In contrast, the recent edition of Allen (2020) lists 43 for Kent, and Pontin (2005) lists 46 in Surrey. Ireland has about 20 in total, plus some species listed as introduced (Niechoj 2011).

Although recording effort is a little greater in the south than the north of Britain and Ireland (Fig. 4.2), it is most

Palaearctic

A biogeographical region that encompasses Asia north of the Himalayas, all of Europe, North Africa and the northern part of the Arabian Peninsula.

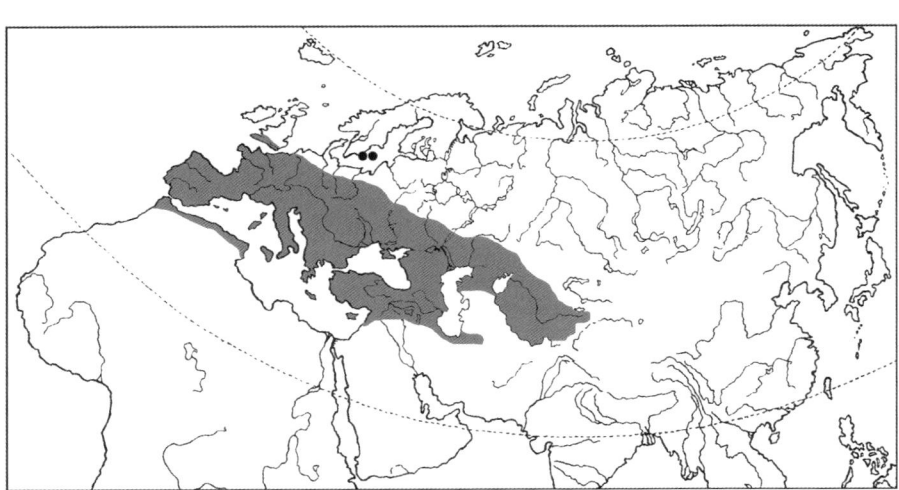

Fig. 4.1 Distribution of *Tapinoma erraticum* in the Palaearctic (from Radchenko 2016)

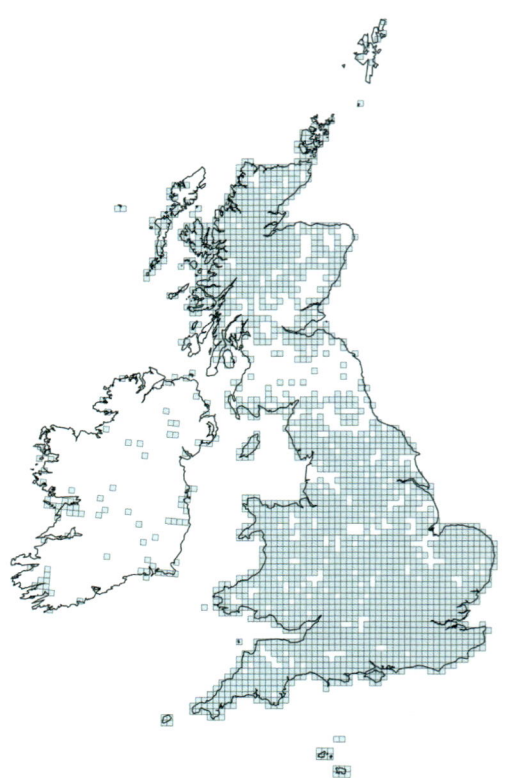

Fig. 4.2 The distribution of ant records received by the BWARS team

likely that these species numbers reflect real ecological requirements. Thus, there tend to be more ants and more species of ants in warmer areas, such as the south-west of the region, and in warmer habitats, such as sandy soils (Fig. 4.3).

4.1 Distribution and species notes

The species listed by the Bees, Wasps and Ants Recording Society (BWARS) has been taken as the definitive species list in this book. This covers Britain, Ireland and associated islands. A brief outline of the biology of each species is provided, together with the distribution maps from BWARS and photographs of specimens of worker ants, or queens for species with no workers. All of the photographs in this chapter are of mounted museum specimens. This means they are in poses that do not mirror real life, but do allow easy comparison between species. In a very few cases the colours are not as in life or body parts are missing or displaced;

distribution maps
We use the maps from BWARS. Other possible sources are the National Biodiversity Network (NBN) Atlas (p. 304), iRecord (p. 299), and county and regional guides (p. 305). Note that some records on maps from one source may be missing in maps from other sources.

vice-county

A geographical division
of Britain and Ireland
used for the purposes
of biological recording
and other scientific
data-gathering.

Fig. 4.3 Number of species per vice-county and zones where daily sunlight in May is between 6 and 7 hours. Blank areas represent 1 to 10 species, lightly dotted 11 to 19 species, densely dotted 20 to 29 species and black over 30 species (from Brian 1977).

this is noted where relevant. The photographs represent the general appearance of the species and, along with the maps, provide a useful overall visual check for identifications made using the keys in Chapter 6. They are not intended to be illustrative of identification features in the keys, such as hairs. The species are treated in alphabetical order.

The maps represent the most up-to-date and comprehensive picture of which ants are where at time of writing. The records are presented over four time periods: pre-1994, 1995–2004, 2005–2014 and 2015–present (a key to time periods is shown to the left). The range of body sizes, excluding legs and antennae, is given at the end of each species account. The sizes are based mainly on Collingwood (1979). Flight periods are also shown (from Lebas *et al.* 2016).

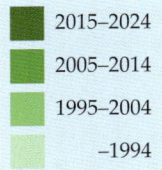

2015–2024

2005–2014

1995–2004

–1994

Key to symbols
on the maps

Aphaenogaster subterranea

This ant is occasionally found on the Channel Islands, where it eats arthropods and tends homopterans. Lebas *et al.* (2016) posit that it feeds petals to its larvae and there is also a suggestion that these are piled up and fungi grow on them, which are then eaten. Its nests are usually quite deep and found under stones or at the base of trees.

In 2024, the presence of a related species was confirmed on Guernsey, the harvester ant *Messor capitatus* (A. Marquis and A. Smith pers. comm.). It is large, black and strongly polymorphic (p. 33). Previously, sightings had been made in 1978 and 1996, so it seems it might be a long-term resident that has been overlooked (despite its size) rather than a casual introduction from nearby France.

Sizes: worker 3.0–5.0 mm; queen 7.0–8.0 mm; male 4.0–4.5 mm

Flight period: July–September

Map 1 *Aphaenogaster subterranea*

Formica aquilonia

This is a mound-building wood ant found in coniferous and birch forest. In favourable locations, it can form huge steep-sided mounds, often larger than those of other wood ants. Its food consists of various prey and scavenged items and honeydew from Homoptera. It locates all this by fanning out from the nest on distinct trails. The nests harbour a wide range of myrmecophiles (including the shining guest ant, *Formicoxenus nitidulus*). Due to its high population densities, *F. aquilonia* can be an ecosystem engineer. All British and Irish wood ants (*F. aquilonia*, *F. lugubris*, *F. pratensis* and *F. rufa*), were generally designated as *Formica rufa* until the revision by Yarrow (1955). Care must be taken when reading papers on wood ants prior to this date. Refer to Maps 7, 9 and 10 for the current known distribution of *F. lugubris, F. pratensis* and *F. rufa*.

Sizes: worker 4.0–9.0 mm; queen 8.0–10.0 mm; male 8.0–10.0 mm

Flight period: May–July

Map 2 *Formica aquilonia*

Formica cunicularia

Found in hot, open areas, this species does not tolerate shade. It is most commonly found on the coast, heaths and downs. It is often seen on flowers of plants in the carrot family (Apiaceae), where it hunts prey and probably takes nectar from the flowers. It is also known to tend aphids and even possibly those that dwell underground. Ants showing the features of a red thorax and legs (var. *rubescens*) and/or red on the head (var. *glebaria*) were long thought of as varieties of *F. fusca*, until Yarrow revised the genus (Yarrow 1954). Older literature therefore needs to be checked for information under the old varietal names of *F. fusca*. The nests harbour a range of myrmecophiles, including the woodlouse *Platyarthrus hoffmannseggii*.

In the specimen depicted here, the right antenna is missing three segments.

Sizes: worker 4.0–7.0 mm; queen 7.5–9.0 mm; male 8.0–9.0 mm

Flight period: June–July

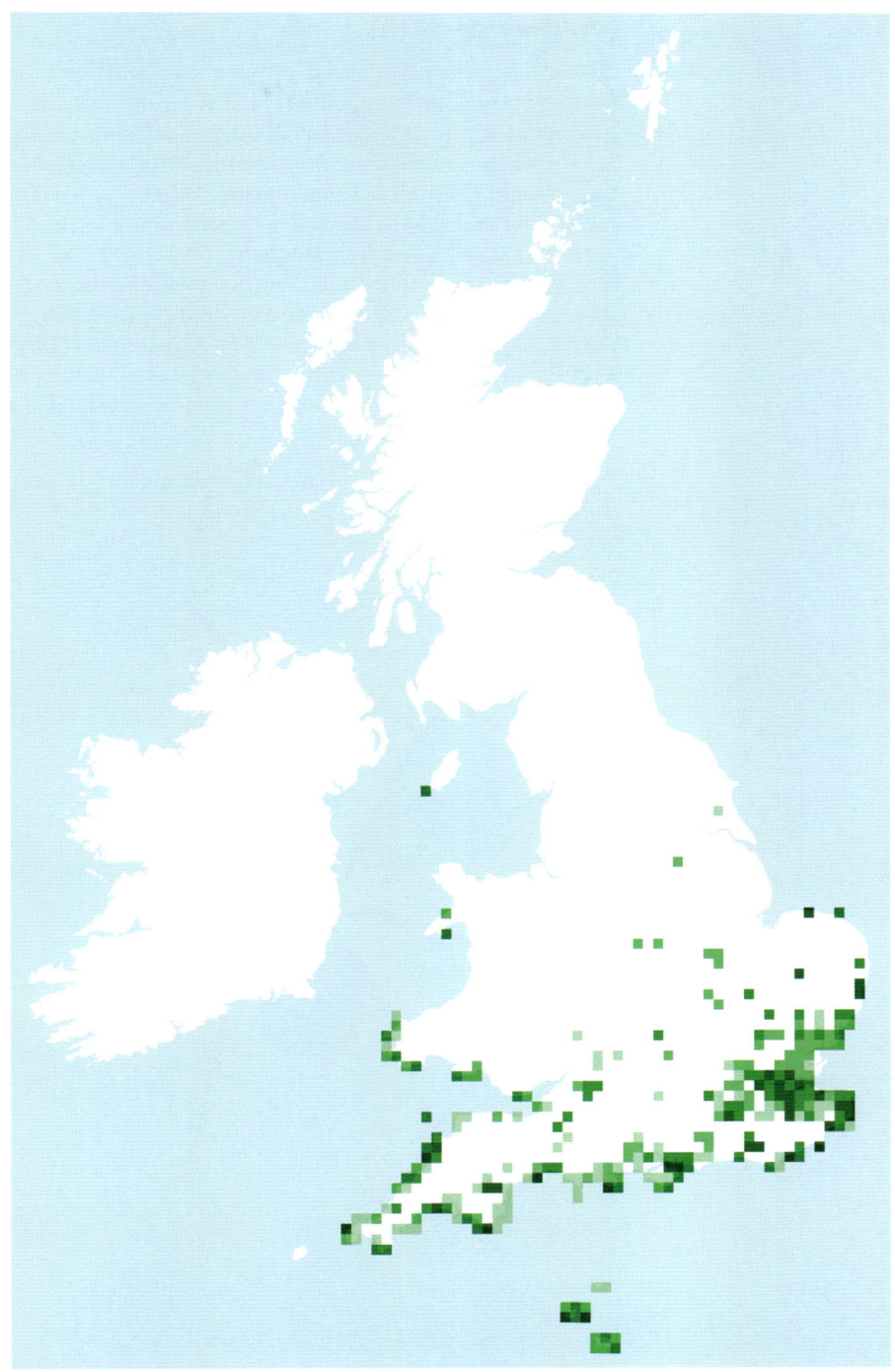

Map 3 *Formica cunicularia*

Formica exsecta

This ant forms small mounds of plant fragments in open habitats such as heath and moor, roadsides, and woodland clearings. Despite its appearance and the thatched mound, *F. exsecta* is not a wood ant; it is a member of the subgenus *Coptoformica* (as opposed to subgenus *Formica*, the wood ants), all of which have a concave region at the back of the head (the occiput). The ant forages in all directions from the nest, with no trails formed. Its diet is like that of the wood ants, consisting of honeydew and all kinds of invertebrate prey. It harbours the wide range of myrmecophiles found with *Formica* species. *F. exsecta* is now a rare ant in England, with just one population surviving in Devon. Scotland has locally strong populations.

Sizes: worker 4.0–7.5 mm; queen 7.5–9.5 mm; male 6.0–9.0 mm

Flight period: July–August

Map 4 *Formica exsecta*

Formica fusca

One of our commonest ants, *F. fusca* is found in open woodland, heaths and other uncultivated habitats. It feeds on invertebrate prey and honeydew but does not form trails. Older sources lump this species with *F. lemani* as the two were not separated until 1954 (Yarrow 1954).

Sizes: worker 4.5–7.5 mm; queen 7.0–9.5 mm; male 8.0–9.5 mm

Flight period: June–July

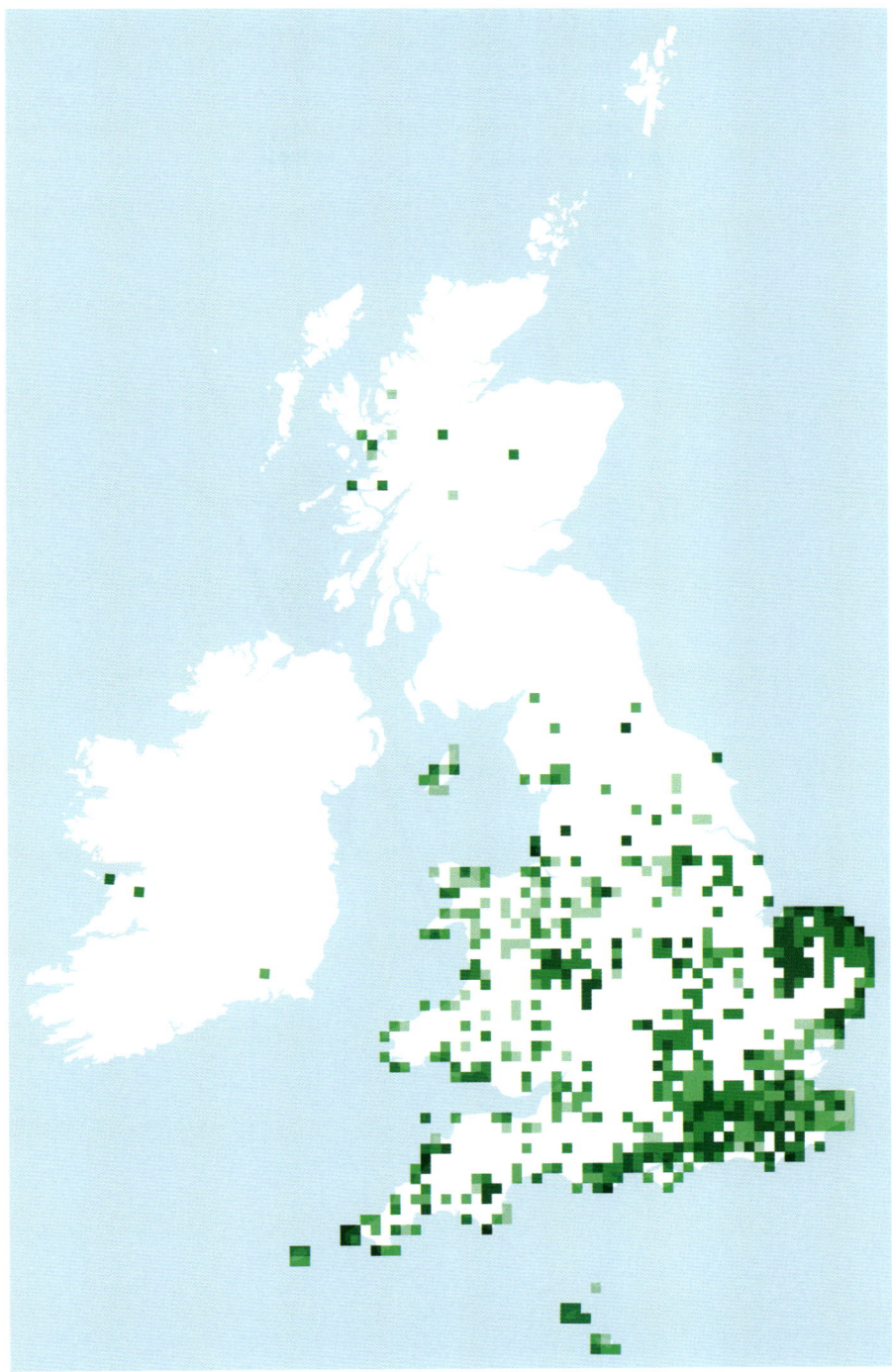

Map 5 *Formica fusca*

Formica lemani

Another very common species. It occupies similar habitats to its close relative *F. fusca*, which it replaces further north and at higher altitudes. The main foods are invertebrates and honeydew.

Sizes: worker 4.5–6.5 mm; queen 7.0–9.5 mm; male 8.0–9.0 mm

Flight period: June–July

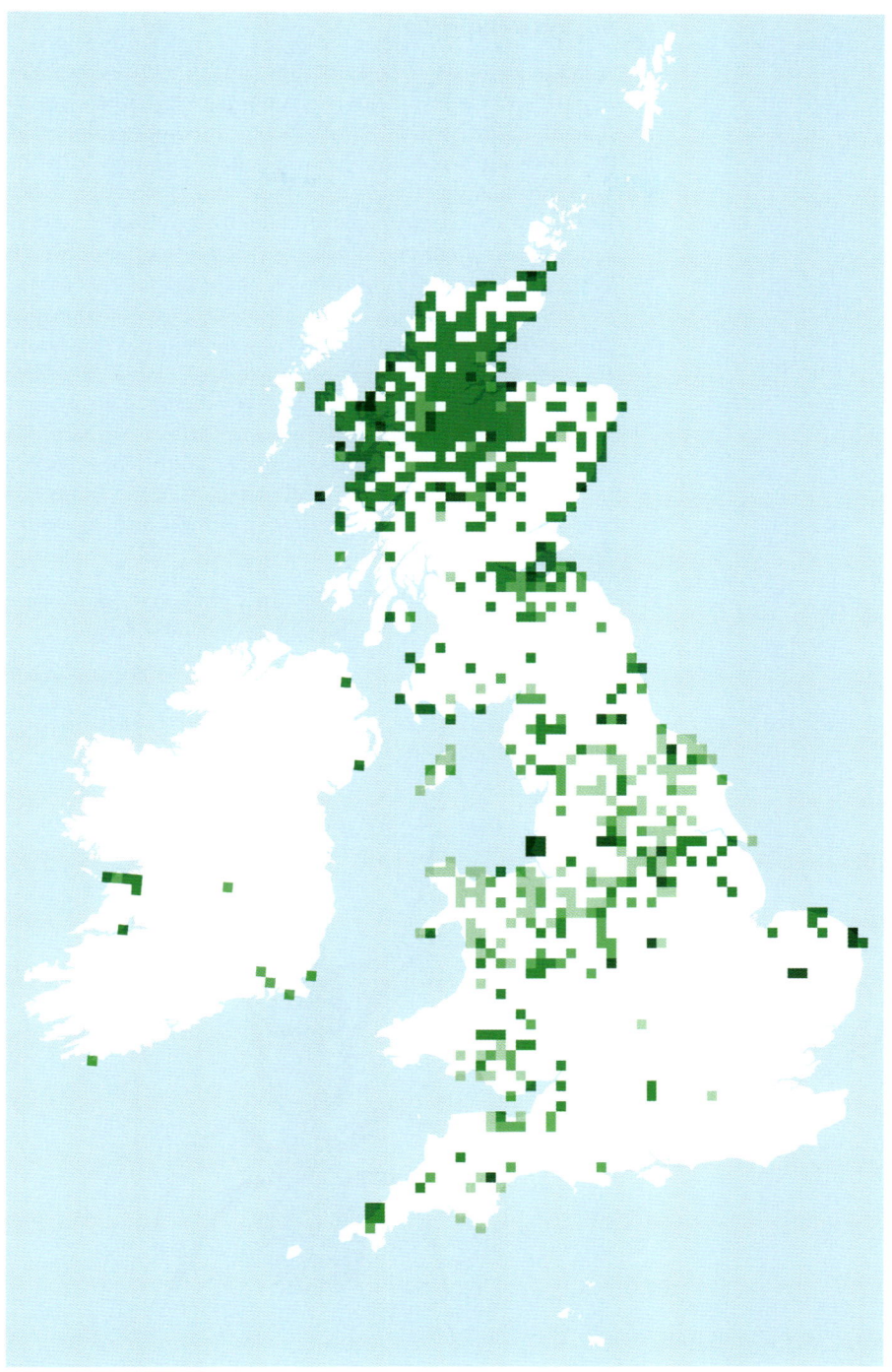

Map 6 *Formica lemani*

Formica lugubris

A mound-building wood ant, sometimes known as the hairy or northern wood ant. This northerly species occupies similar habitats to those of *F. rufa*, although it is more tolerant of exposure and seems to colonise plantations more easily. As with the other wood ants, it feeds on animal material, which is either killed or scavenged, and honeydew. Like the others too, it forms distinct trails along which foraging occurs. It harbours many myrmecophiles, including *Formicoxenus nitidulus*. Again, in common with the other wood ants, this species is declining, and this may have some important consequences due to its ecosystem-engineer status. All British and Irish wood ants (*F. aquilonia*, *F. lugubris*, *F. pratensis* and *F. rufa*), were generally designated as *Formica rufa* until the revision by Yarrow (1955). Care must be taken when reading papers on wood ants prior to this date. Refer to Maps 2, 9 and 10 for the current known distribution of *F. aquilonia*, *F. pratensis* and *F. rufa*.

In the specimen depicted here, the right foreleg is missing its tarsus and the right maxillary palp is detached.

Sizes: worker 4.0–9.0 mm; queen 9.5–10.5 mm; male 9.5–10.5 mm

Flight period: May–July

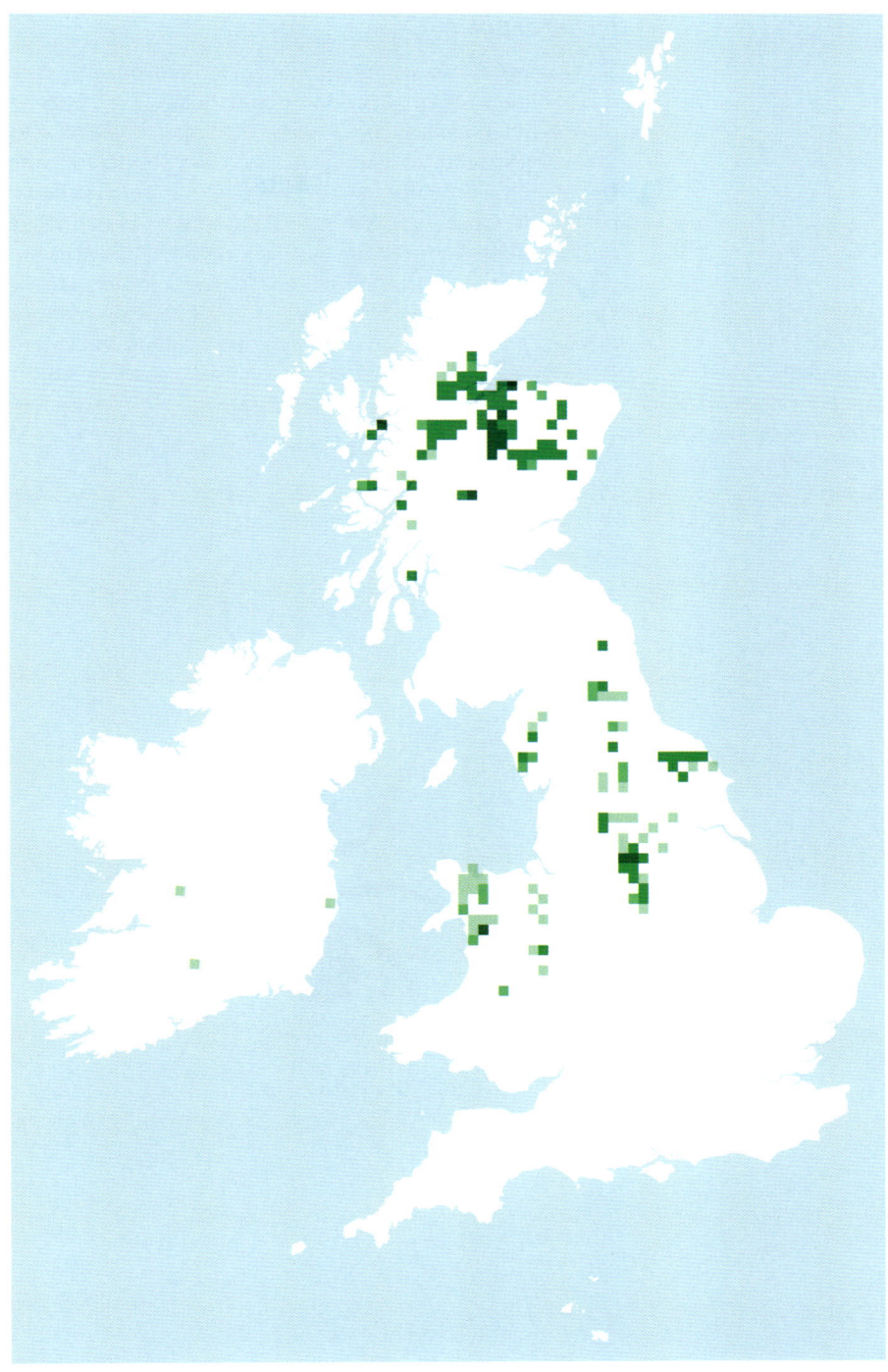

Map 7 *Formica lugubris*

Formica picea

This rare species is unique in Britain in its preference for very wet habitats such as bogs. Some populations have only been discovered in the last 20–30 years, a fact most likely due to the inaccessibility of its habitat. One such population was discovered as recently as 2003, so exploration of suitable habitat elsewhere may well reveal this rare species. It relies heavily on honeydew from aphids for food. Some prey are returned to the nest and it is known to 'steal' insects caught by sundew plants. Workers forage singly and no trails are formed. The workers can endure complete submergence and can also survive being supercooled. Little work has been done on myrmecophiles of this species.

supercool

In supercooling, a liquid cools below its freezing point without changing phase into a solid

Sizes: worker 4.0–5.5 mm; queen 8.0–9.0 mm; male 7.5–8.5 mm

Flight period: July–August

Map 8 *Formica picea*

Formica pratensis

This species seems to be less dependent on woodland than the other wood ants. It is found in scrub on heathland and, in the Channel Islands, on cliffs near coastal paths. It feeds on honeydew and collects prey, all along distinct trails. Common on the Channel Islands, it used to be found around Bournemouth but has not been seen on the mainland since 1987. The list of myrmecophiles is short, but this may be due to its own rarity in the region. It does include *Formicoxenus nitidulus* in continental Europe, so would be worth looking for on the Channel Islands. All British and Irish wood ants (*F. aquilonia*, *F. lugubris*, *F. pratensis* and *F. rufa*), were generally designated as *Formica rufa* until the revision by Yarrow (1955). Care must be taken when reading papers on wood ants prior to this date. Refer to Maps 2, 7 and 10 for the current known distribution of *F. aquilonia*, *F. lugubris* and *F. rufa*.

Sizes: worker 4.0–9.0 mm; queen 9.5–11.5 mm; male 9.5–11.5 mm

Flight period: May–August

Map 9 *Formica pratensis*

Formica rufa

A mound-building wood ant found in lowland forest. It tends to favour more open areas, such as the edges of rides and in glades. Its food consists of various prey and scavenged items and honeydew from Homoptera. It locates all this by fanning out from the nest on distinct trails. The nests harbour a wide range of myrmecophiles (including the shining guest ant, *Formicoxenus nitidulus*). Due to its high numbers, *F. rufa* can be an ecosystem engineer and its relative decline may therefore have wide-reaching consequences for associated animals, plants and nutrient cycles. For example, nest-building creates new microhabitats (Subedi 2016). All British and Irish wood ants were designated as *Formica rufa* until the revision by Yarrow (1955). Care must be taken when reading papers on wood ants prior to this date. Refer to the maps of the distribution of these species as now known. Refer to maps 2, 7 and 9 for the current known distribution of *F. aquilonia*, *F. lugubris* and *F. pratensis*. It is now known that the ant found in Britain that has been known for centuries as *Formica rufa* is a stable hybrid between *Formica polyctena* and *F. rufa* (p. 36).

In the specimen depicted here, the left antenna is missing all but two of the segments of the funiculus.

Sizes: worker 4.0–9.0 mm; queen 9.5–11.0 mm; male 9.0–11.0 mm

Flight period: April–May

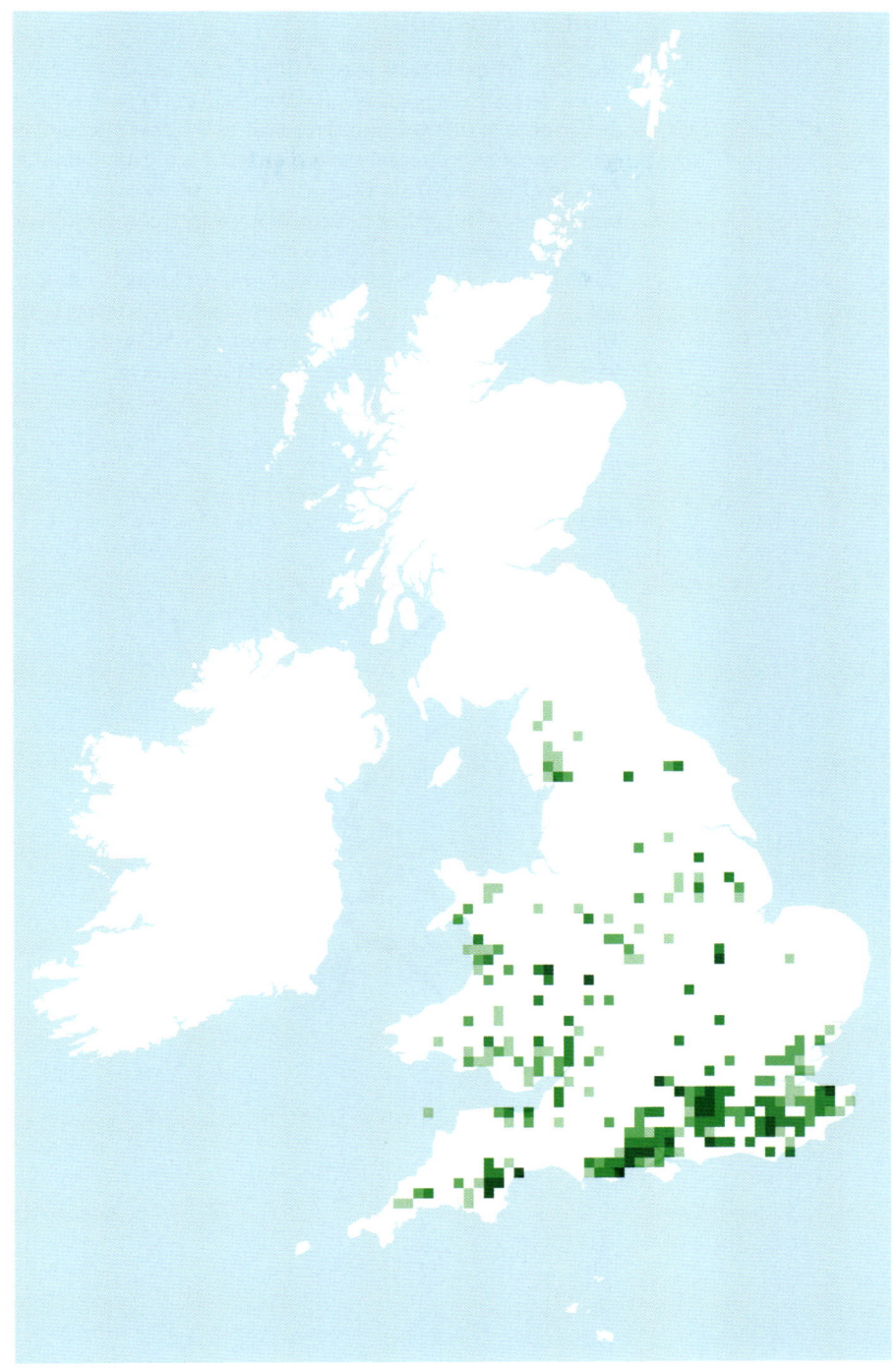

Map 10 *Formica rufa*

Formica rufibarbis

This rare species seems to favour hot sites, being generally found on south-facing banks in heaths. It is very tolerant of desiccation. It forages singly for prey and honeydew and is also known to commonly take nectar, like the very similar *F. cunicularia*. Workers often cooperate in the carrying of larger prey. The ant is extremely rare in the region. It is thought to have died out at the previously well-established sites in Surrey that are shown on the map. However, there is one recently discovered site in Hampshire where it survives. This species is quite abundant on one of the Isles of Scilly. It remains vulnerable to further habitat loss, as well as being heavily raided by *Formica sanguinea*.

Sizes: worker 4.5–7.5 mm; queen 8.5–10.5 mm; male 8.5–9.5 mm

Flight period: June–July

Map 11 *Formica rufibarbis*

Formica sanguinea

This is a slave-making species, sometimes called the blood-red slave-maker. Its raiding behaviour is described on p. 69. It is mainly found in sandy soils, often with woodland, and usually pine. It has a striking disjointed distribution in Britain (see Map 12). Foraging is generally by single workers, although a trail may be formed to large prey or when on a slave raid. There is a range of myrmecophiles.

Sizes: worker 6.0–9.0 mm; queen 9.0–11.0 mm; male 7.0–10.0 mm

Flight period: July

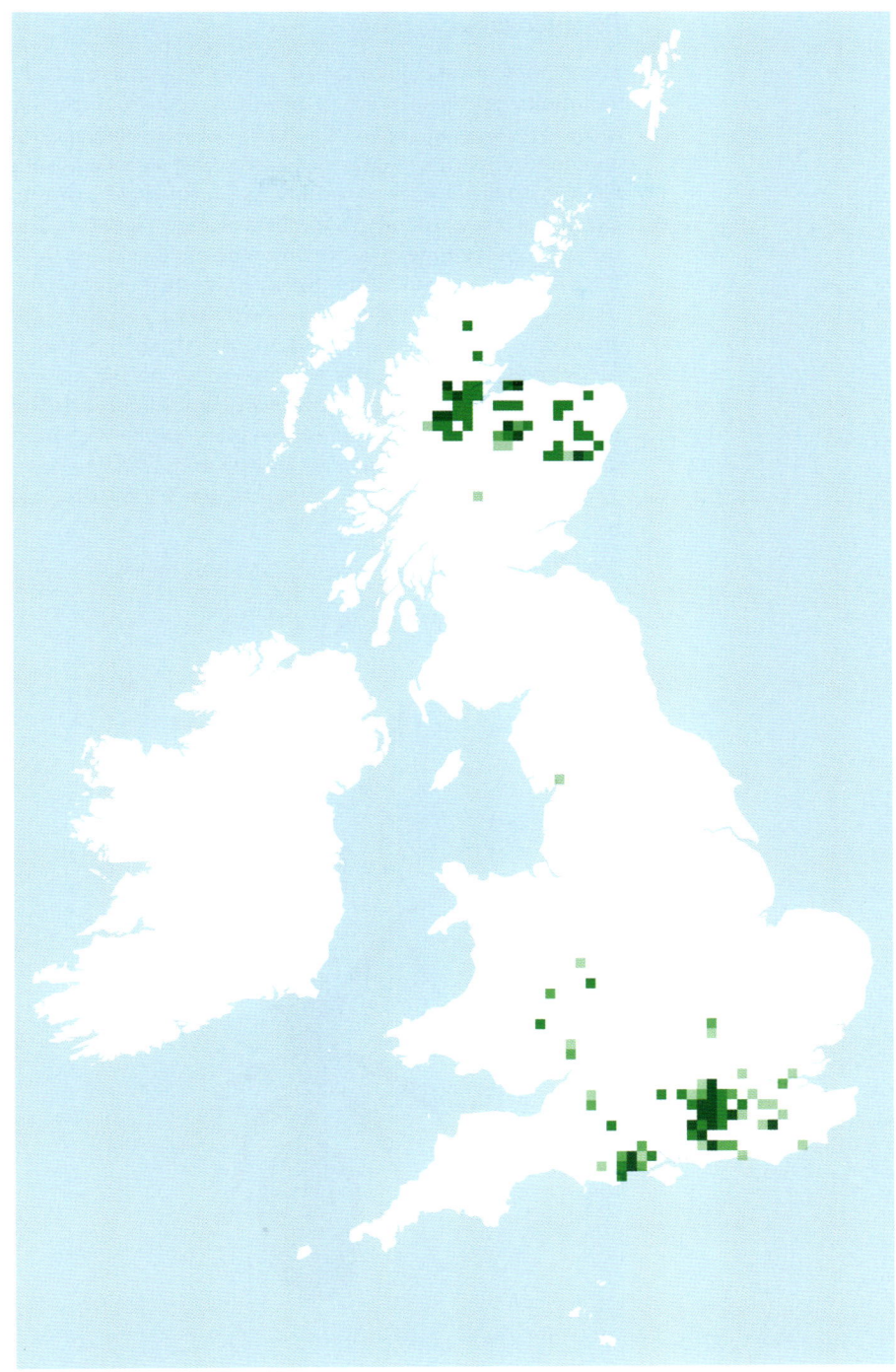

Map 12 *Formica sanguinea*

Formicoxenus nitidulus

The shining guest ant is one-of-a-kind in Britain as our only xenobiotic species (p. 67). It is widely distributed, with records from all the areas where its host wood-ant species are found, although it has not been recorded in Ireland, with either *F. aquilonia* in the north or with *F. lugubris* in the south. Usually, it is only found in a few host nests at a location. The difficulty of spotting it, as its tiny workers spend much of their time in the wood-ant mound, suggests it may well be under-recorded. It should be looked for wherever there are wood ants, the best times being August–October, at which time the worker-like males are very active on the mound surface, or on warm humid days at any time of the year. The males cannot fly, and mating takes place on the wood-ant mound. It is thought to feed mainly on and from host ants. The specimen shown in the photograph is a gyne (queen); workers look very similar.

Sizes: worker 2.8–3.4 mm; queen 3.4–3.6 mm; male 2.8–3.2 mm

Flight period: August

Map 13 *Formicoxenus nitidulus*

Hypoponera ergatandria

sibling species
When two or more species are identical or nearly identical in their appearance, yet reproductively isolated.

Found in buildings, such as hothouses, in Britain and very occasionally outside. It nests in the ground, in logs and under bark in moist conditions. It feeds on small invertebrates such as Collembola. Map 14 shows records that are definitely *H. ergatandria* as opposed to its sibling species *H. punctatissima* (see also Maps 15 and 62).

Sizes: worker 2.5–3.2 mm; queen 3.4–3.6 mm; male 3.4–3.6 mm

Flight period: August–February

Map 14 *Hypoponera ergatandria*

Hypoponera punctatissima

A cosmopolitan species, which is included on the list due to a few colonies that have been found outside, albeit in areas warmed by fermentation, such as compost heaps. Like *H. ergatandria*, it nests in the ground and feeds on small invertebrates, especially Collembola. Map 15 shows *H. punctatissima* records that are either from a period after the separation of *H. ergatandria* or old records that have been re-examined and shown to be definitely *H. punctatissima*. See Map 62 (p. 204) for older records that have not been re-examined, plus newer records where the recorder has not been able to confidently separate the two species and therefore the ant recorded could have been either *H. punctatissima* or *H. ergatandria*.

Sizes: worker 2.5–3.2 mm; queen 3.4–3.6 mm; male 3.4–3.6 mm

Flight period: August–September

Map 15 *Hypoponera punctatissima*

Lasius alienus

One of a pair of sibling species, the other being *Lasius psammophilus*. It is found in chalk grasslands, on coasts and in similar habitats where it is warm. It depends predominantly on honeydew for food but does consume invertebrates in the spring. Map 16 shows *L. alienus* records that are either from a period after the separation of *L. psammophilus* or old records that have been re-examined and shown to be definitely *L. alienus*. See Map 63 (p. 205) for older records that have not been re-examined, plus newer records where the recorder has not been able to confidently separate the two species and therefore the ant recorded could have been either *L. alienus* or *L. psammophilus*.

Sizes: worker 3.0–4.2 mm; queen 8.0–9.0 mm; male 3.0–3.8 mm

Flight period: July–August

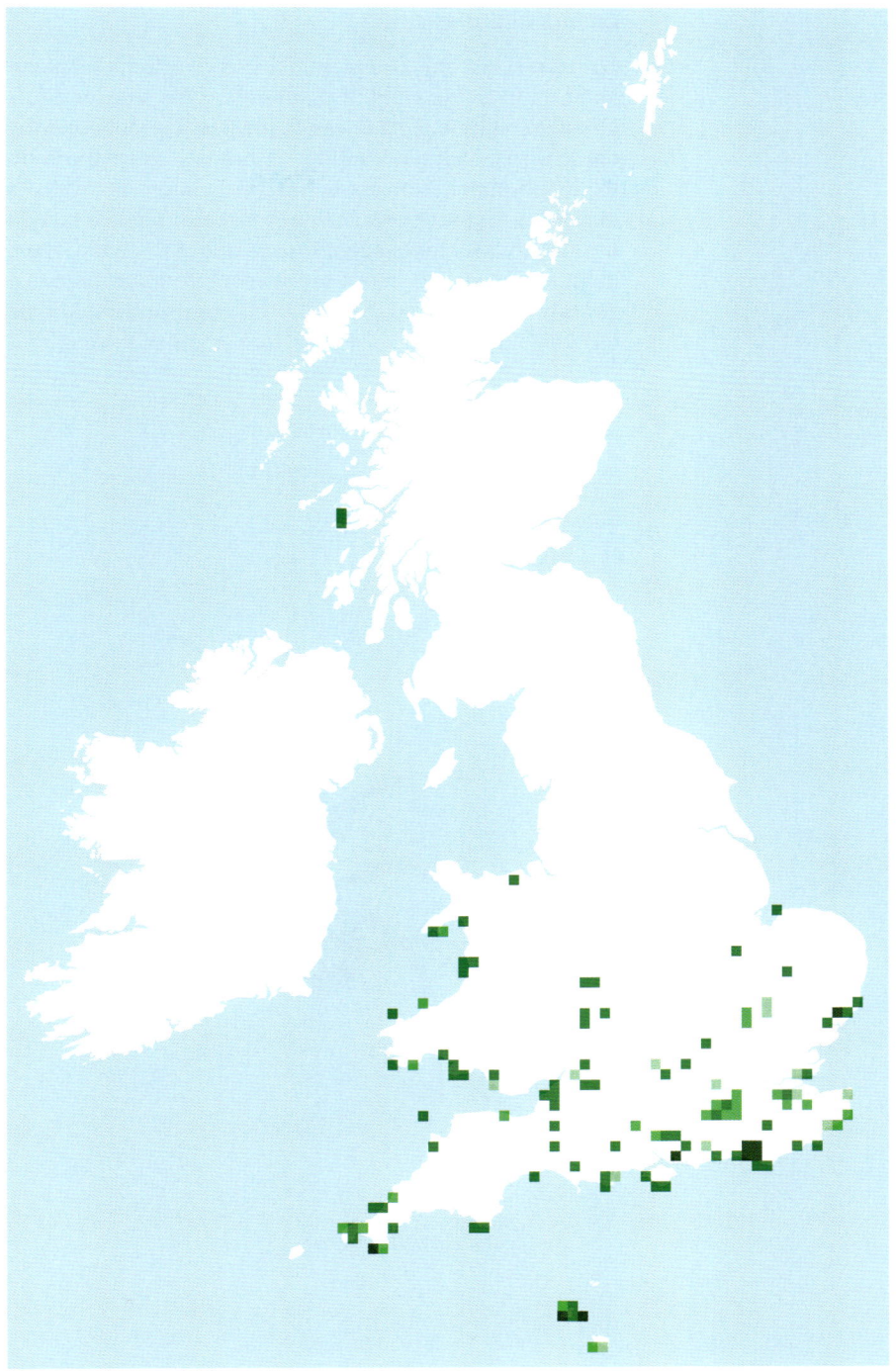

Map 16 *Lasius alienus*

Lasius brunneus

The brown tree ant is quite unique in the region for making its nests in living trees. It is often overlooked due to its nest being within the tree, often down in the roots, and the timid workers largely confined to tunnels under the bark. Its distribution has increased markedly in recent years. Its main food seems to be honeydew from bark aphids, although it probably also takes small insects under the bark. There is little information about myrmecophiles, but there is a suggestion it is associated with the green-underside blue butterfly, *Glaucopsyche alexis*, which is found throughout continental Europe. This butterfly is not resident in Britain, although it was once caught in Torquay, Devon. It may be worth keeping an eye out for it.

Sizes: worker 3.2–4.5 mm; queen 8.0–9.0 mm; male 3.5–4.5 mm

Flight period: June–July

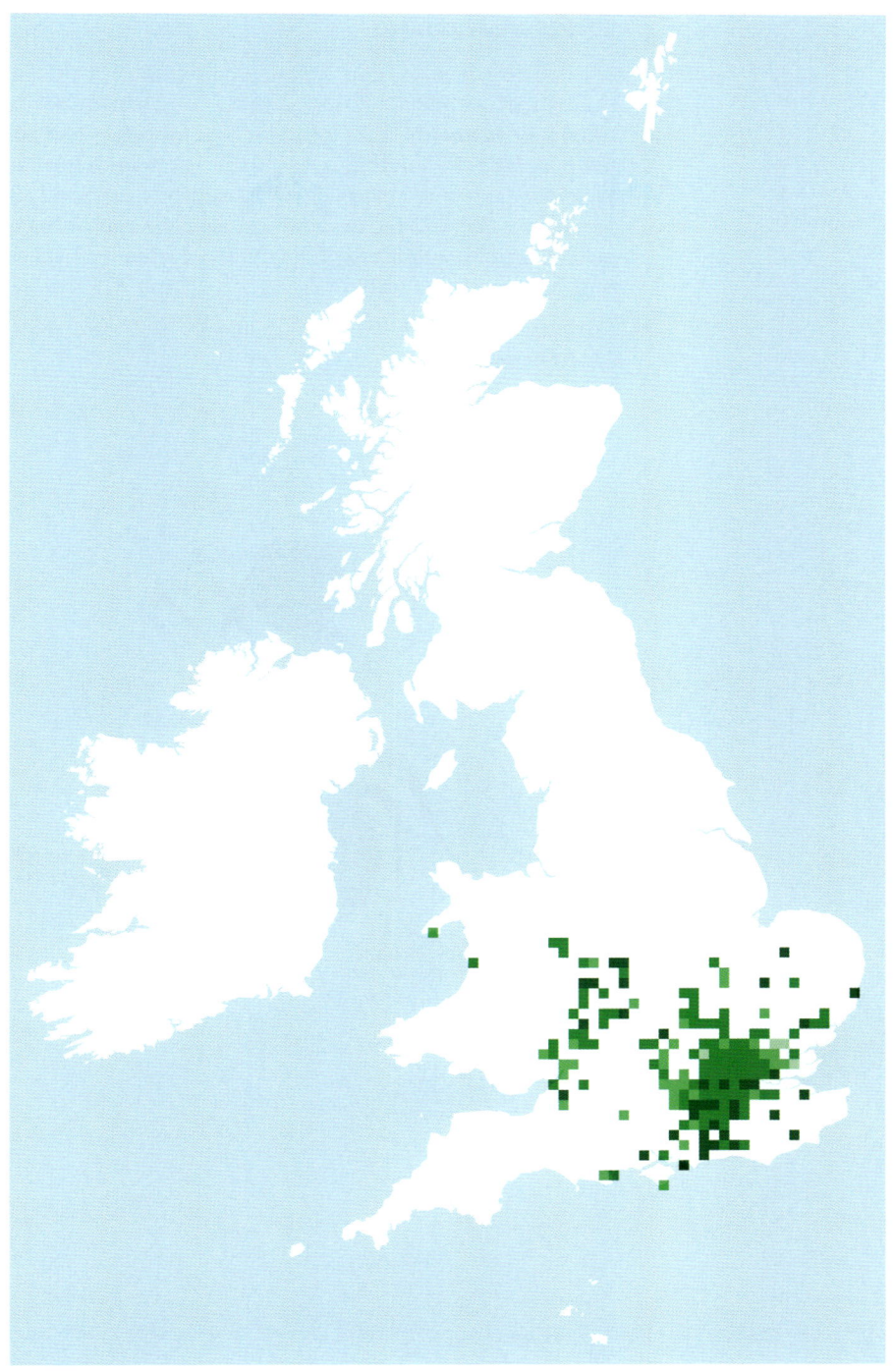

Map 17 *Lasius brunneus*

Lasius emarginatus

A distinctly bicoloured, warmth-loving species that is common in continental Europe and has been known on the Channel Islands for many years. It was found in 2008 on the mainland, in London, and has since spread to further locations, as yet unmapped, in the south of England. It would be worth looking in suitable warm, especially urban, habitats in this general region for more occurrences. It takes prey and a lot of honeydew.

Sizes: worker 3.0–5.5 mm; queen 7.0–10.0 mm; male 7.0–14.0 mm

Flight period: July–August

Map 18 *Lasius emarginatus*

Lasius flavus

The yellow hill ant or yellow meadow ant is one of our best-known species due to the conspicuous mounds it may build in undisturbed grasslands and pastures. It does not, however, always build mounds and can be found in a wide range of habitats. Workers are rarely seen as this species is largely confined to a subterranean existence. Here it tends root aphids for honeydew and eats small soil insects. In less-disturbed habitats, nest density is very high and *L. flavus* is a potent ecosystem engineer, having effects on both soil chemistry and vegetation. The extensive nests harbour a wide range of myrmecophiles and the woodlouse *Platyarthrus hoffmannseggii* can be particularly conspicuous.

Sizes: worker 2.2–4.8 mm; queen 7.2–9.5 mm; male 3.5–5.0 mm

Flight period: August–October

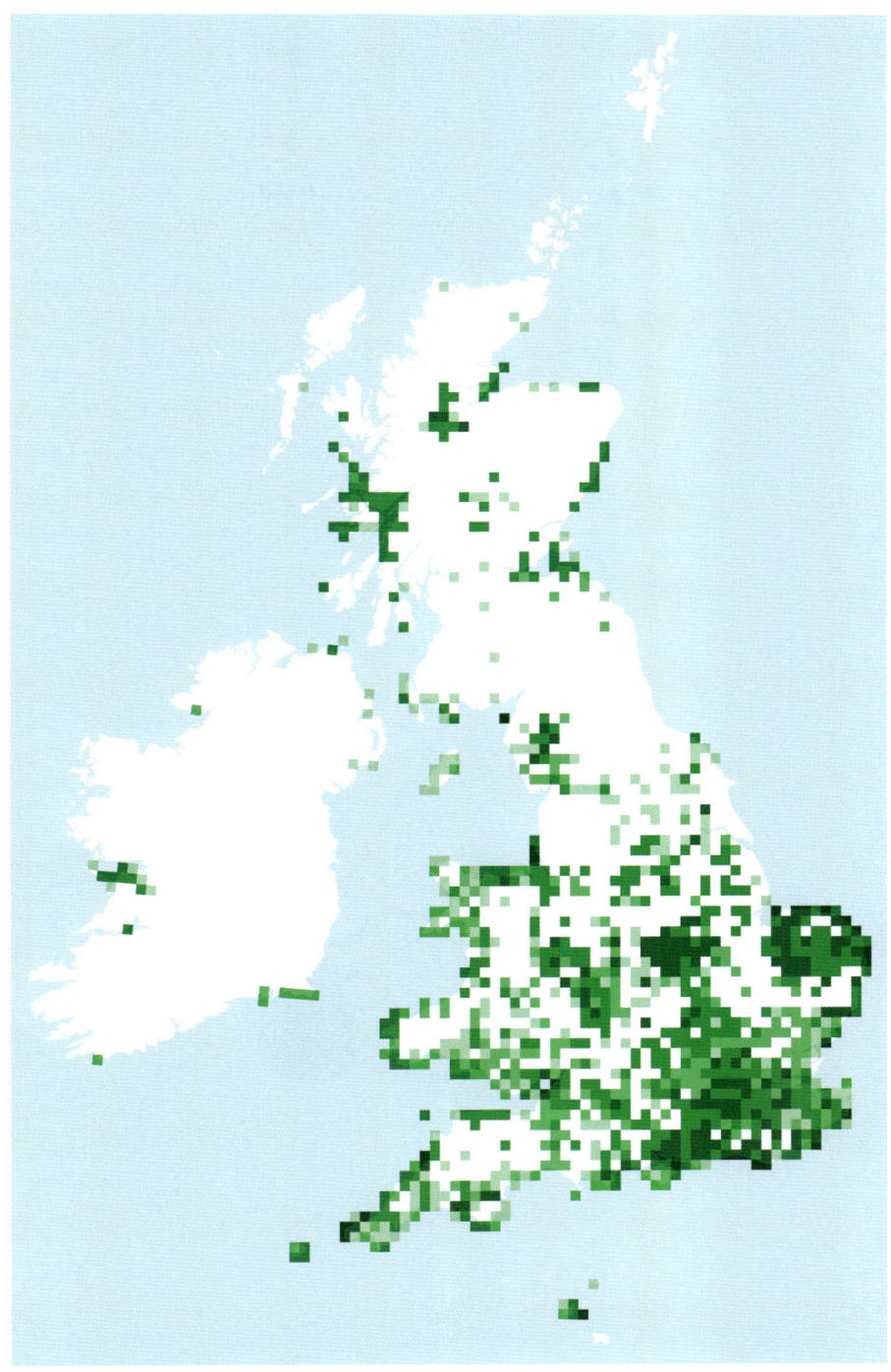

Map 19 *Lasius flavus*

Lasius fuliginosus

The jet-black ant is one of our most distinctive species. With its shiny body, notched (heart-shaped) head and strong smell of oranges it cannot be mistaken for anything else. It is a temporary social parasite (p. 67) of *Lasius* species related to *L. umbratus* (which themselves are also social parasites). *L. fuliginosus* is thus a social hyperparasite. Nests are often in dead trees or stumps. It forms very distinctive busy trails to plants where it tends Homoptera for honeydew, its main food. It is also seen carrying dead insects, often aphids, back to the nest. A wide range of myrmecophiles has been found in the nest, which is made of a material called carton, consisting of chewed wood fragments mixed with honeydew.

Sizes: worker 4.0–6.0 mm; queen 6.0–6.5 mm; male 4.5–5.0 mm

Flight period: May–September

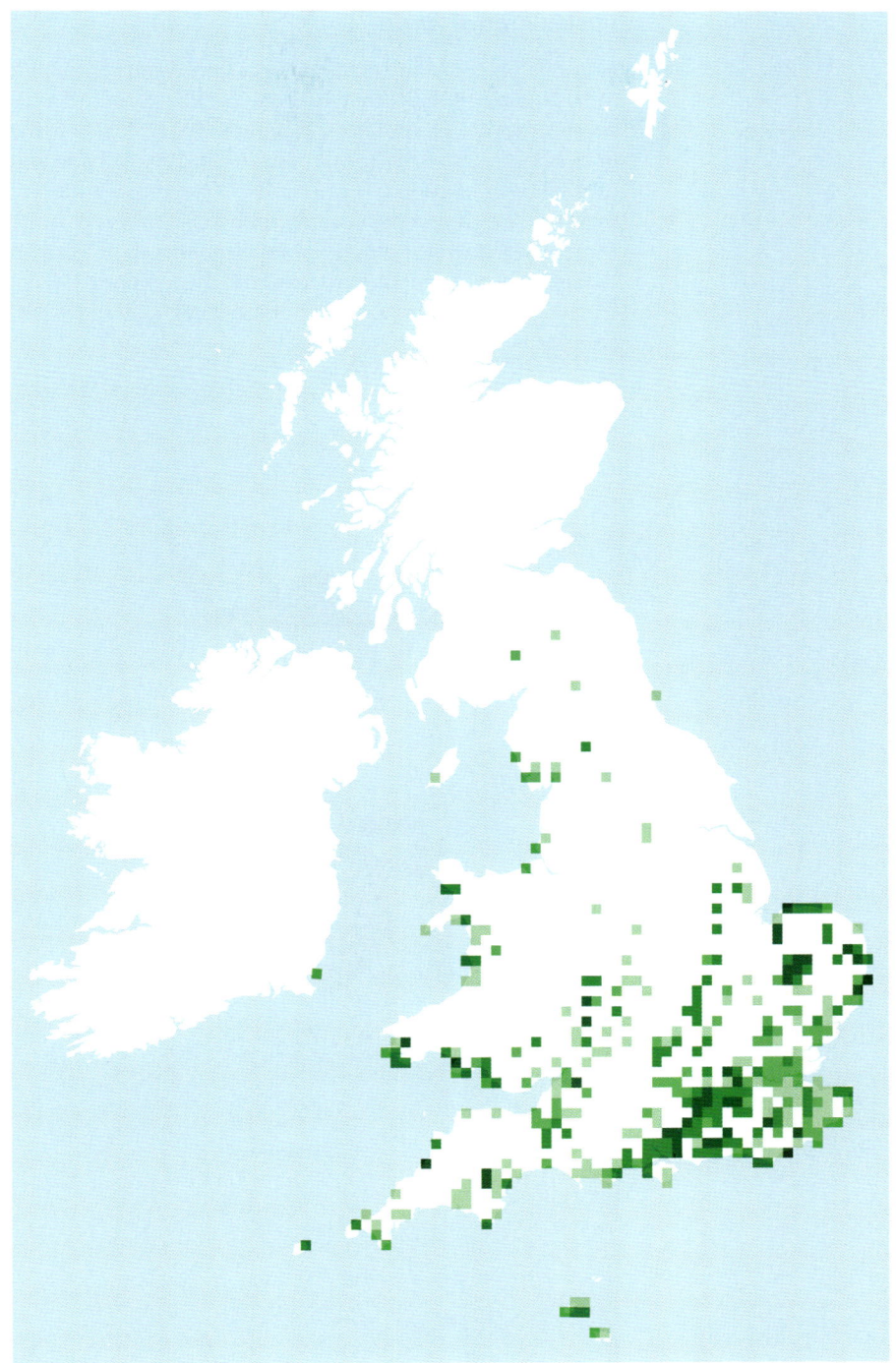

Map 20 *Lasius fuliginosus*

Lasius meridionalis

Found in warm, sandy areas, such as dunes and heathland, this is quite a rare species. It is one of the group of *Lasius* species that are temporary social parasites, in this case the host being *L. psammophilus*. In fact, it (and related species) may be most frequently recorded by observing wandering gynes in search of a host colony to invade. Nests are quite rarely seen and not much is known about their general biology. They are thought to feed mainly on honeydew. In the specimen shown here, the terminal segment of the left antenna is missing.

Sizes: worker 3.5–5.0 mm; queen 7.0–8.0 mm; male 4.0–4.5 mm

Flight period: May–October

Map 21 *Lasius meridionalis*

Lasius mixtus

Found in pastures and thin woodland, this species does not require the level of warmth sought by many other *Lasius* species. It is a temporary social parasite of *Lasius flavus*. Newly mated gynes hibernate before seeking out host nests in early spring. Its feeding is little studied, but it is probably mainly dependent on underground aphids. Donisthorpe (1927) does not mention this species but lists a wide range of myrmecophiles for the closely related *L. umbratus*, including *Platyarthrus hoffmannseggii*. This ant is brighter yellow in life.

Sizes: worker 3.5–4.5 mm; queen 6.0–7.5 mm; male 4.2–4.8 mm

Flight period: May–October

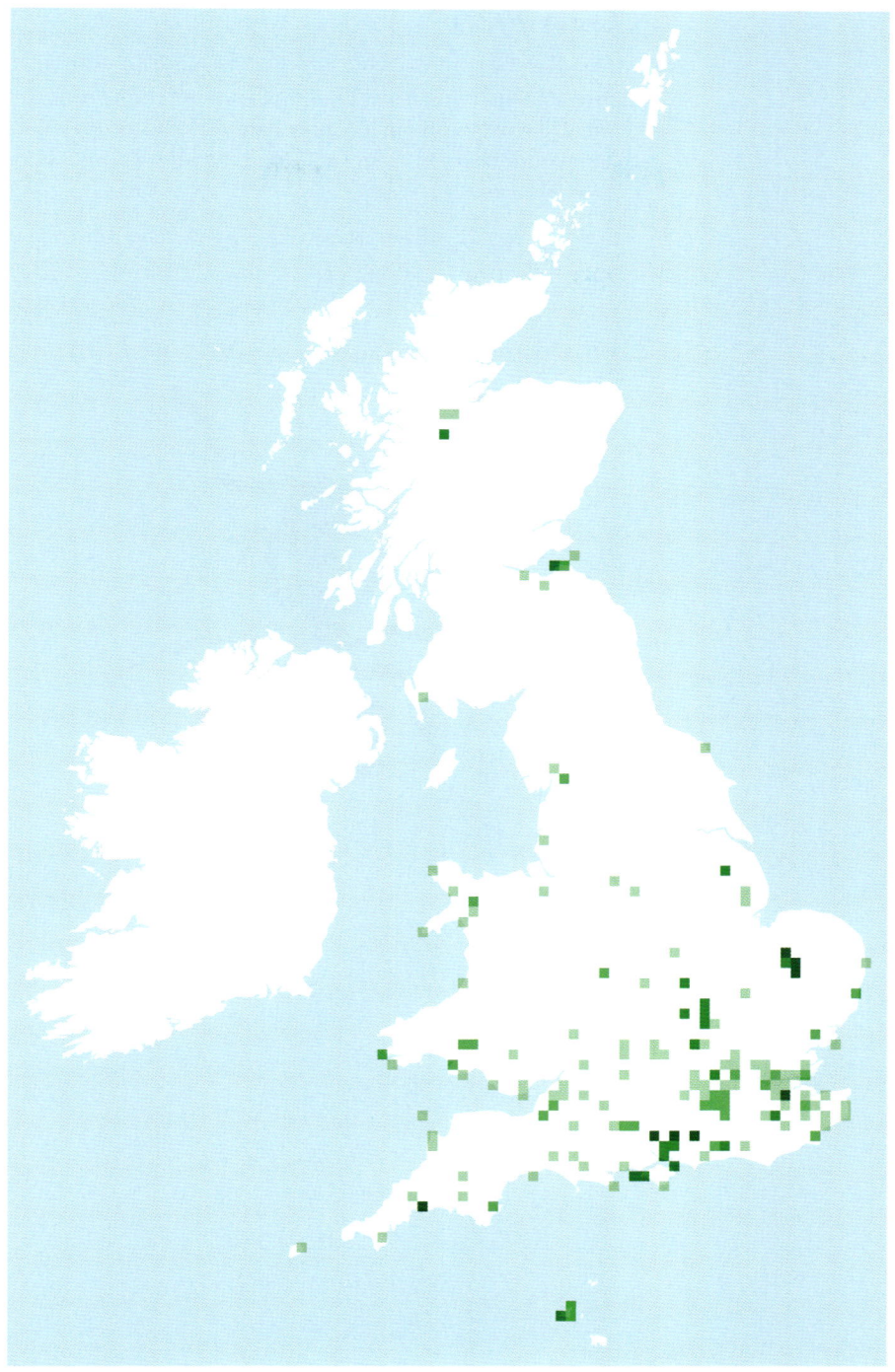

Map 22 *Lasius mixtus*

Lasius myops

There are a few records from the Channel Islands. It is closely related to *L. flavus* but is not known to build mounds and exclusively prefers hot, dry habitats. It feeds on aphid honeydew, derived from root-living species, and possibly invertebrates.

Sizes: worker 2.2–4.8 mm; queen 7.2–9.5 mm; male 3.5–5.0 mm

Flight period: August–October

Map 23 *Lasius myops*

Lasius neglectus

The invasive garden ant was first seen in Britain in 2009 at Hidcote Manor in Gloucestershire. Since then, it has been found at a few further sites, one being as far north as Kirk Smeaton in Yorkshire. All the sites are urban or managed, such as parks and gardens. It probably originates from Asia Minor but, since its appearance in Budapest in the mid-1970s, it is now known from 19 European countries. It feeds mainly on honeydew. Supercolonies are formed, which are colonies containing numerous queens, and these can effectively eliminate all other ants from the area. This ant may well be overlooked, due to its similarity to native species, until it has reached pest proportions.

In the specimen depicted here, the right foreleg is folded under, as is the tarsus of the left foreleg; the rear leg on the left is missing its tarsus.

Sizes: worker 2.5–3.0 mm; queen 5.5–6.0 mm; male 2.0–3.0 mm

Flight period: not known to have a nuptial flight as such, but sexuals are in the nest from June to August. However, queens can fly.

Map 24 *Lasius neglectus*

Lasius niger

The black garden ant or common black ant is surely the species most people are familiar with, due to its propensity to nest in urban areas and enter houses in search of sweet foods. In the wild they catch prey and tend homopterans. It is known to protect Homoptera colonies and is a very aggressive species, quite prone to bite. In addition to taking sugar solution from homopterans, it is also often seen feeding at extrafloral nectaries. Map 25 shows *L. niger* records that are either from a period after the separation of *L. platythorax* or old records that have been re-examined and shown to be *L. niger*. See Map 64 (p. 206) for older records that have not been re-examined, plus newer records where the recorder has not been able to confidently separate the two species and therefore the ant recorded could have been either *L. niger* or *L. platythorax*.

Sizes: worker 3.5–5.0 mm; queen 8.0–9.0 mm; male 3.5–5.0 mm

Flight period: July–August

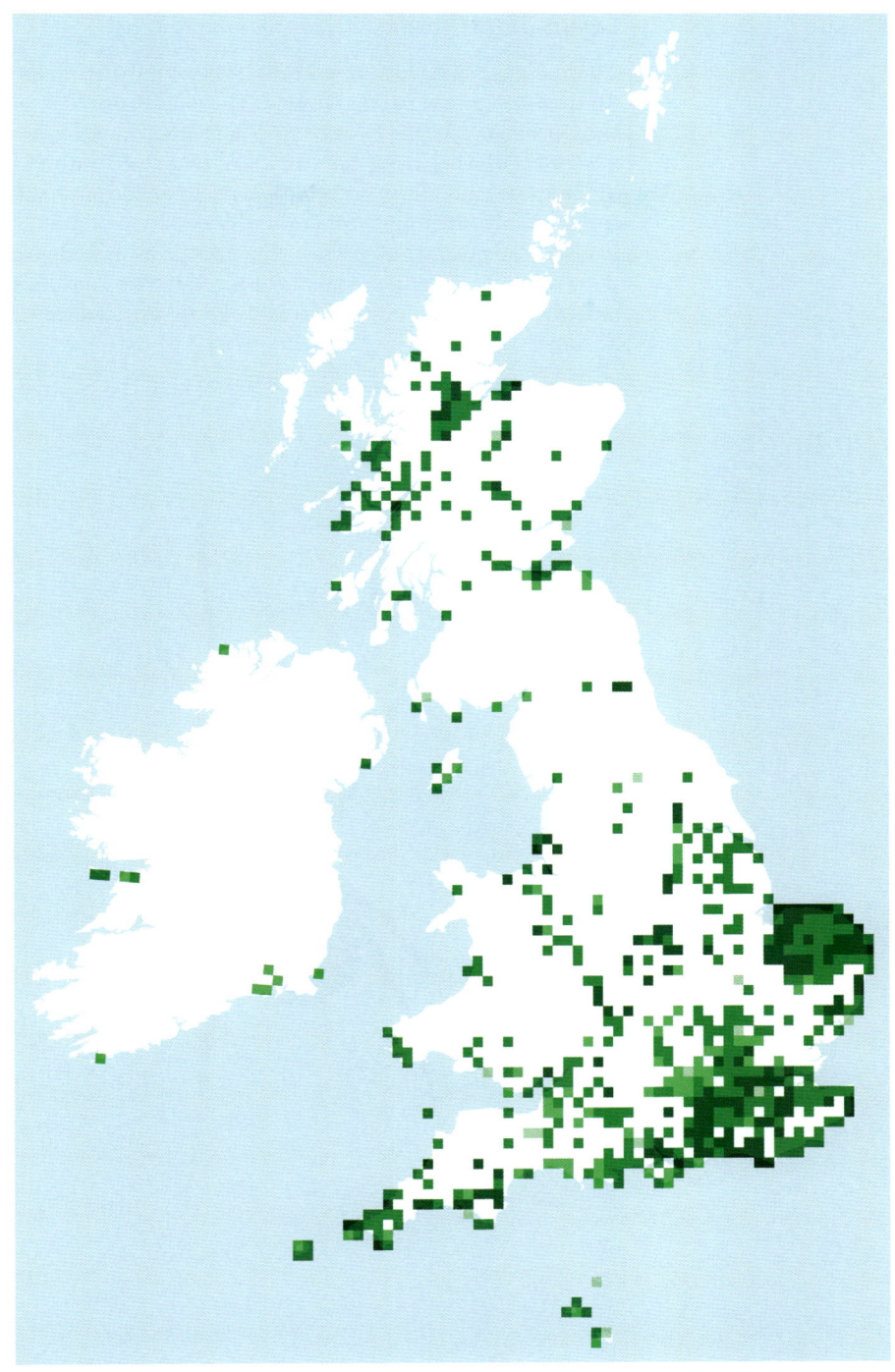

Map 25 *Lasius niger*

Lasius platythorax

Confused for a long time with its close relative *L. niger*, *L. platythorax* was separated in 1991. The microscopic differences are accompanied by very different habitat preferences. *L. niger* avoids cool, damp habitats, including woodland, and it is in these situations that *L. platythorax* may be dominant. *L. platythorax* often nests in and under dead wood, unlike *L. niger*. An aggressive species, which feeds as a predator, but also tends Homoptera and takes nectar. Map 26 shows records that are definitely *L. platythorax* as opposed to its sibling *L. niger* (see also Maps 25 and 64)

Sizes: worker 3.0–5.0 mm; queen 9.0 mm; male 3.5–5.0 mm

Flight period: July–August

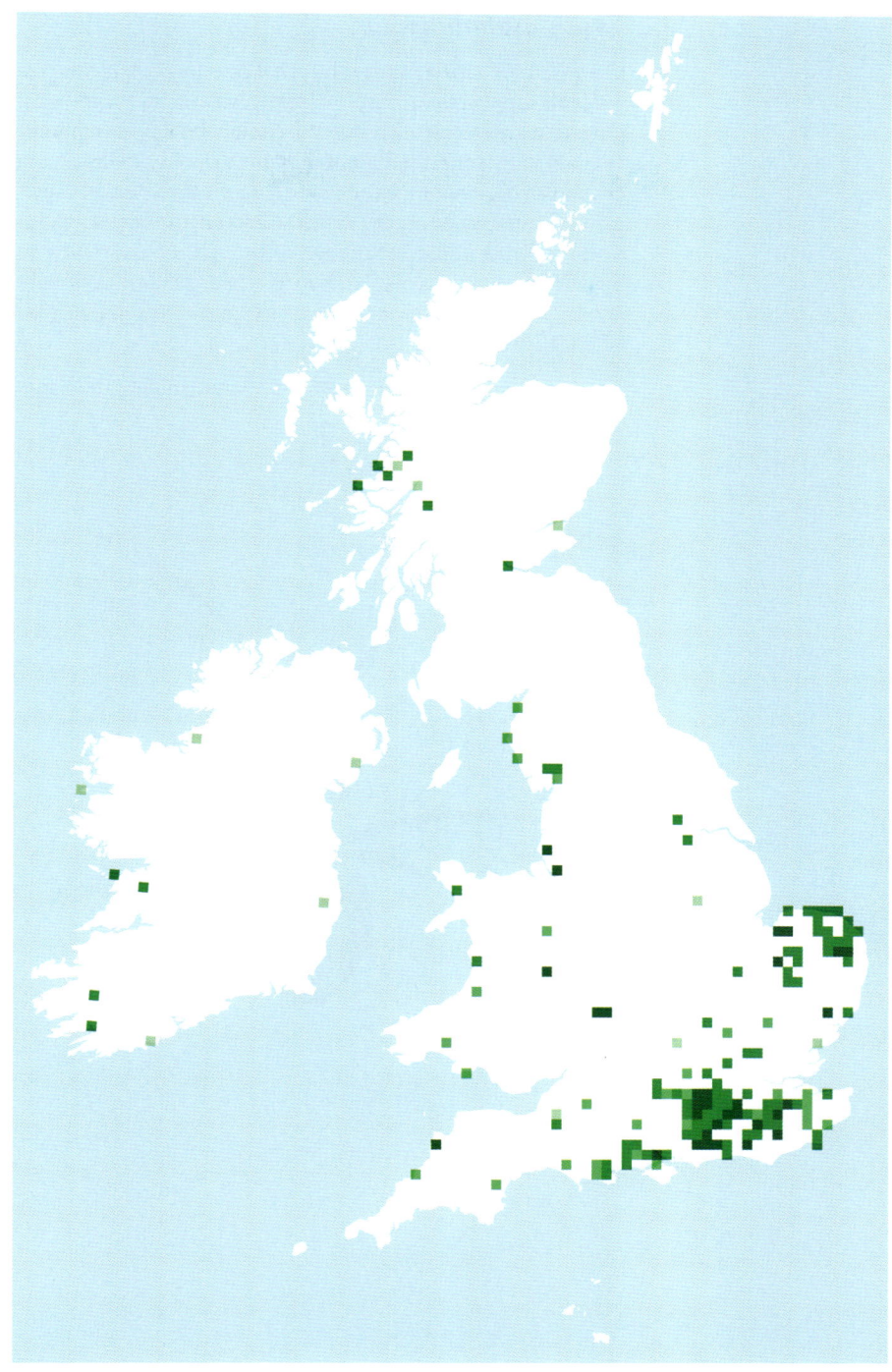

Map 26 *Lasius platythorax*

Lasius psammophilus

Confused for a long time with *L. alienus*, *L. psammophilus* was accorded species status in 1992. Microscopic morphological differences go along with very different habitat requirements. It tends to be found in heathland and in sandy areas on the coast. Its food is a mixture of prey and above- and below-ground homopterans, but much worker activity takes place underground. Due to confusion with *L. alienus*, it is likely to be under-recorded. Map 27 shows records that are definitely *L. psammophilus* as opposed to its sibling *L. alienus* (see also Maps 16 and 63).

Sizes: worker 3.0–4.2 mm; queen 8.0–9.0 mm; male 3.0–3.8 mm

Flight period: July–August

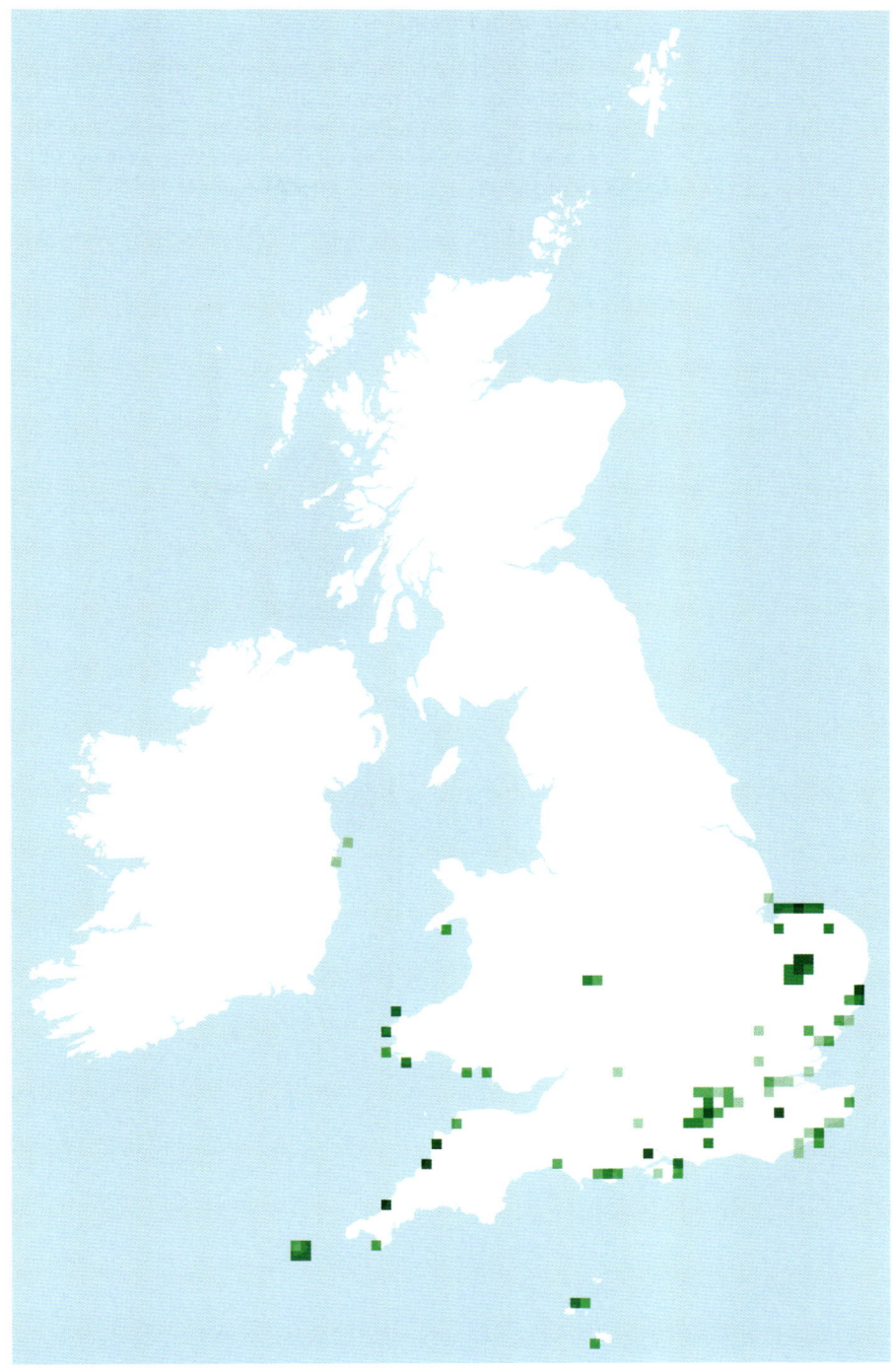

Map 27 *Lasius psammophilus*

Lasius sabularum

Very similar to *L. mixtus*, the species was clarified as separate in 1988. It is found in parks and gardens and is thought to feed on honeydew from root-feeding Homoptera. Like *L. mixtus*, this species is a temporary social parasite. Its hosts may include *L. niger* and related species.

Sizes: worker 3.5–5.0 mm; queen 7.0–8.0 mm; male 4.0–4.5 mm

Flight period: May–October

Map 28 *Lasius sabularum*

Lasius umbratus

This temporary social parasite is found in the habitats in which its hosts live (*L. niger*, *L. platythorax*, *L. brunneus* and *L. psammophilus*). Its nests are often associated with dead wood and can include carton constructed by the ants, similar to the nests of *L. fuliginosus*. Workers spend most of their time underground feeding on honeydew from homopterans and predating small invertebrates.

Sizes: worker 3.5–5.0 mm; queen 7.0–8.0 mm; male 4.0–4.5 mm

Flight period: May–October

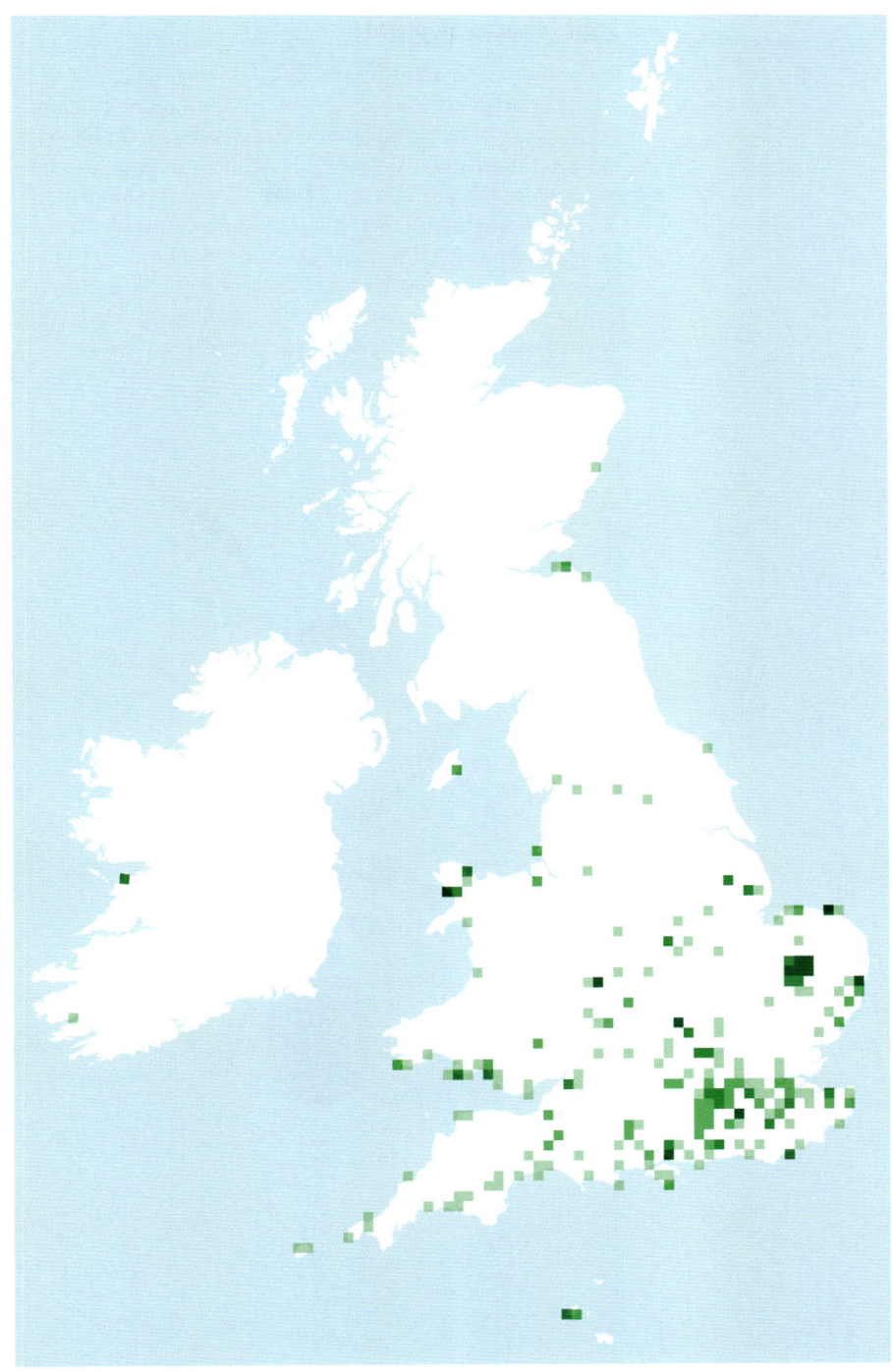

Map 29 *Lasius umbratus*

Leptothorax acervorum

A common and widespread ant, but inconspicuous and forming small colonies. Nests in a wide range of habitats but favours open woodland, where it can be found nesting in dead wood and under bark. In other situations, it can be encountered under stones and at the bases of plants such as heather. Foraging for food can take place up to 6 m from the nest, during which it will catch a wide range of arthropods. It forages singly, but a worker can lead another to a food source by so-called tandem recruitment. It will drink honeydew where it is found on surfaces but is not known to tend aphids.

Sizes: worker 3.8–4.5 mm; queen 3.8–4.8 mm; male 4.5–5.0 mm

Flight period: July–September

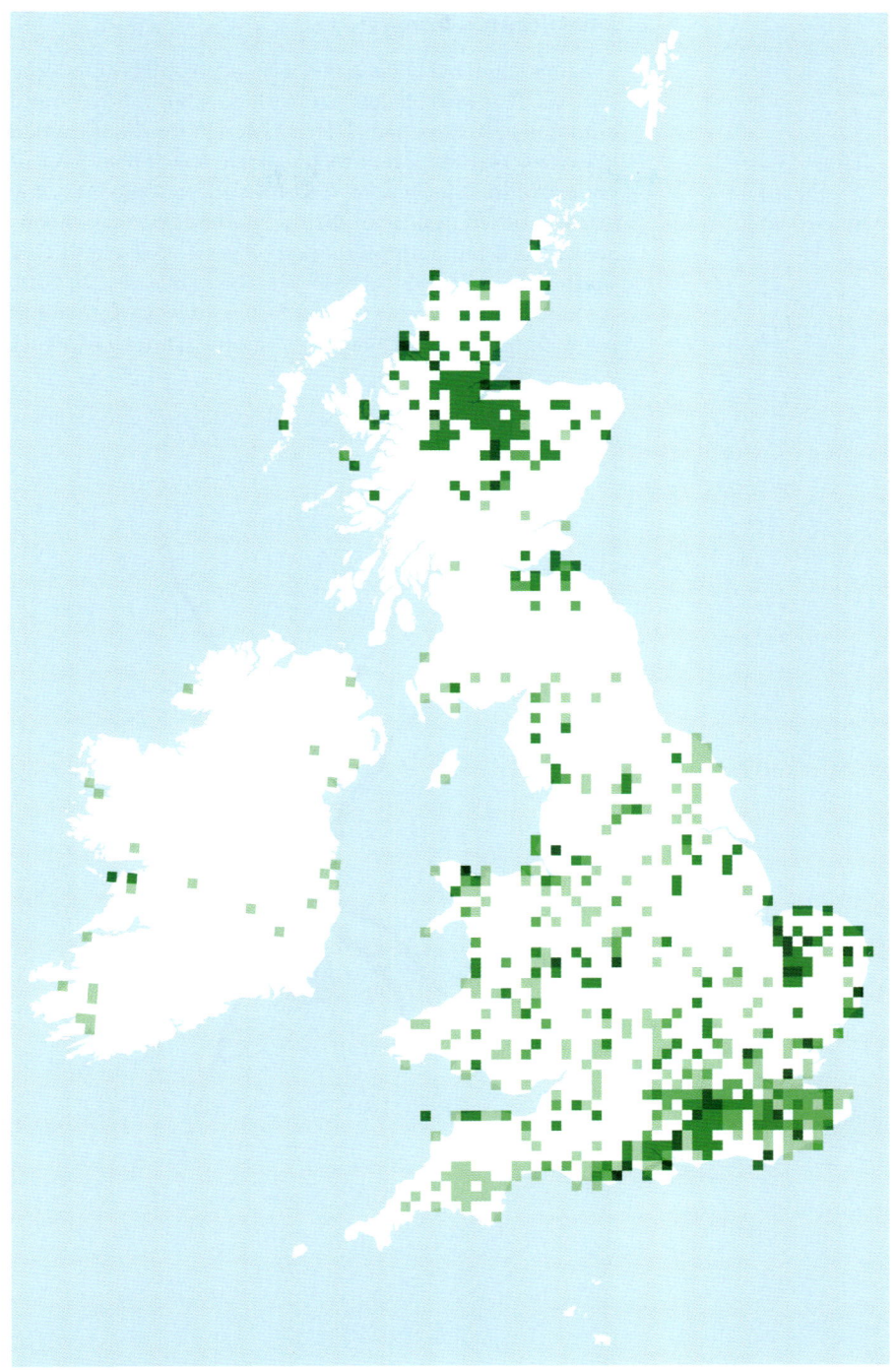

Map 30 *Leptothorax acervorum*

Linepithema humile

The Argentine ant *Linepithema humile* is a notorious introduced species. This species has been, albeit very rarely, found outdoors in Britain, and so is included in the identification keys. It is fortunate that this invasive species, so far, has not established itself here. It forms supercolonies of many interconnected nests (one such in southern Europe extends about 6,000 km along the coast). The workers tend pest species of aphid and compete with the native ants. In 2016, colonies of a close relative *Linepithema iniquum* were found in glasshouses at the National Botanic Garden of Wales (Hamer & Cocks 2016).

Sizes: worker 2.2–2.8 mm; queen 4.5 mm; male 2.5–3.0 mm

Flight period: April–June

Map 31 *Linepithema humile*

Monomorium pharaonis

Colonies of pharaoh ant can be enormous, with many millions of workers which feed on all kinds of scraps. It is found very widely in buildings. These include hospitals, where the species can be a real concern as it can act as a vector for disease. It has been shown to carry *Salmonella*, *Staphylococcus*, *Clostridium* and *Streptococcus* bacteria. In addition, it is very difficult to eradicate, nesting deep within the fabric of the building and having millions of workers and undergoing regular colony fragmentation. It is regarded as a major pest worldwide. The geographical origins of this species seem to be lost in the mists of time. The name is suggestive of an Egyptian origin, coming from the mistaken idea that it constituted one of the plagues that, according to the Bible, afflicted that country.

Sizes: worker 2.0–2.4 mm; queen 4.0–4.8 mm; male 3.0 mm

Flight period: July–August

Map 32 *Monomorium pharaonis*

Myrmecina graminicola

This inconspicuous and slow-moving species forms small nests in a wide range of situations, including cliffs, downland, pastures and gardens, along with open woodland and parks. It is known as the 'woodlouse ant' because of its behaviour of rolling into a ball when disturbed. Unusually, nests may contain individuals with intermediate gyne/worker morphology. The workers forage individually for invertebrate prey and are not known to tend aphids. No myrmecophiles are known.

Sizes: worker 3.0–3.6 mm; queen 4.0–4.2 mm; male 3.4–4.0 mm

Flight period: April–August

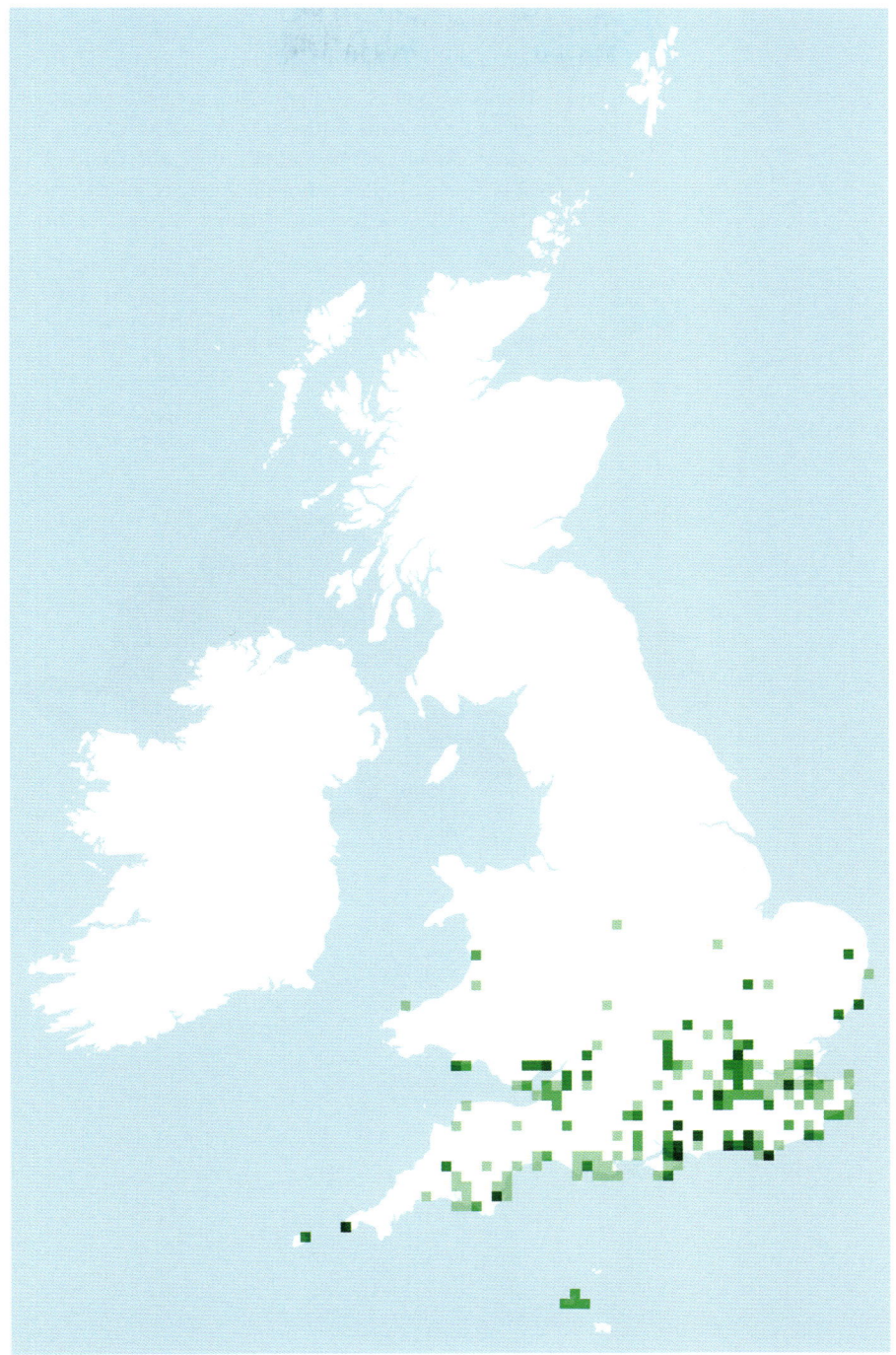

Map 33 *Myrmecina graminicola*

Myrmica hirsuta

A rare parasite found only in nests of its host, *Myrmica sabuleti*. It almost lacks workers (the few present in nests are actually pseudogynes or 'reduced queens'). It is almost certainly seriously under-recorded and would be worth looking for whenever strong populations of the host are encountered. This species' food is all acquired from host workers.

The image depicts a specimen of a gyne or queen, mounted on a point to highlight the extreme hairiness of this species.

Sizes: there are no true workers; queen 4.5 mm; male 5.3 mm

Flight period: July–September

Map 34 *Myrmica hirsuta*

Myrmica karavajevi

This rare workerless parasite of other *Myrmica* species (in Britain *M. scabrinodis* and *M. sabuleti*) used to be placed in a separate genus named *Sifolinia*. The ant is hard to find and is only really evident when gynes and males are produced in the host nest. In an excavation of 11 *M. scabrinodis* colonies, three (27%) were infested, so the species is probably more common than currently thought, especially in locations where its hosts are abundant, such as boggy locations in the south of England. Pitfall trapping has been a useful method for finding this elusive ant (p. 279).

Only the right wing is present in the specimen depicted here, which is a gyne or queen.

Sizes: there are no workers; queen 3.2–3.6 mm; male 3.5 mm

Flight period: July–September

Map 35 *Myrmica karavajevi*

Myrmica lobicornis

A dark-coloured species, most commonly found in upland areas of rough grassland, moorland and open woods. It is widespread, but nests tend to be isolated and inconspicuous. The feeding habits are poorly known, but it may tend Homoptera and otherwise be a scavenger. Only a couple of myrmecophiles have been recorded.

Sizes: worker 4.0–5.0 mm; queen 5.0–5.5 mm; male 5.0–5.5 mm

Flight period: August–September

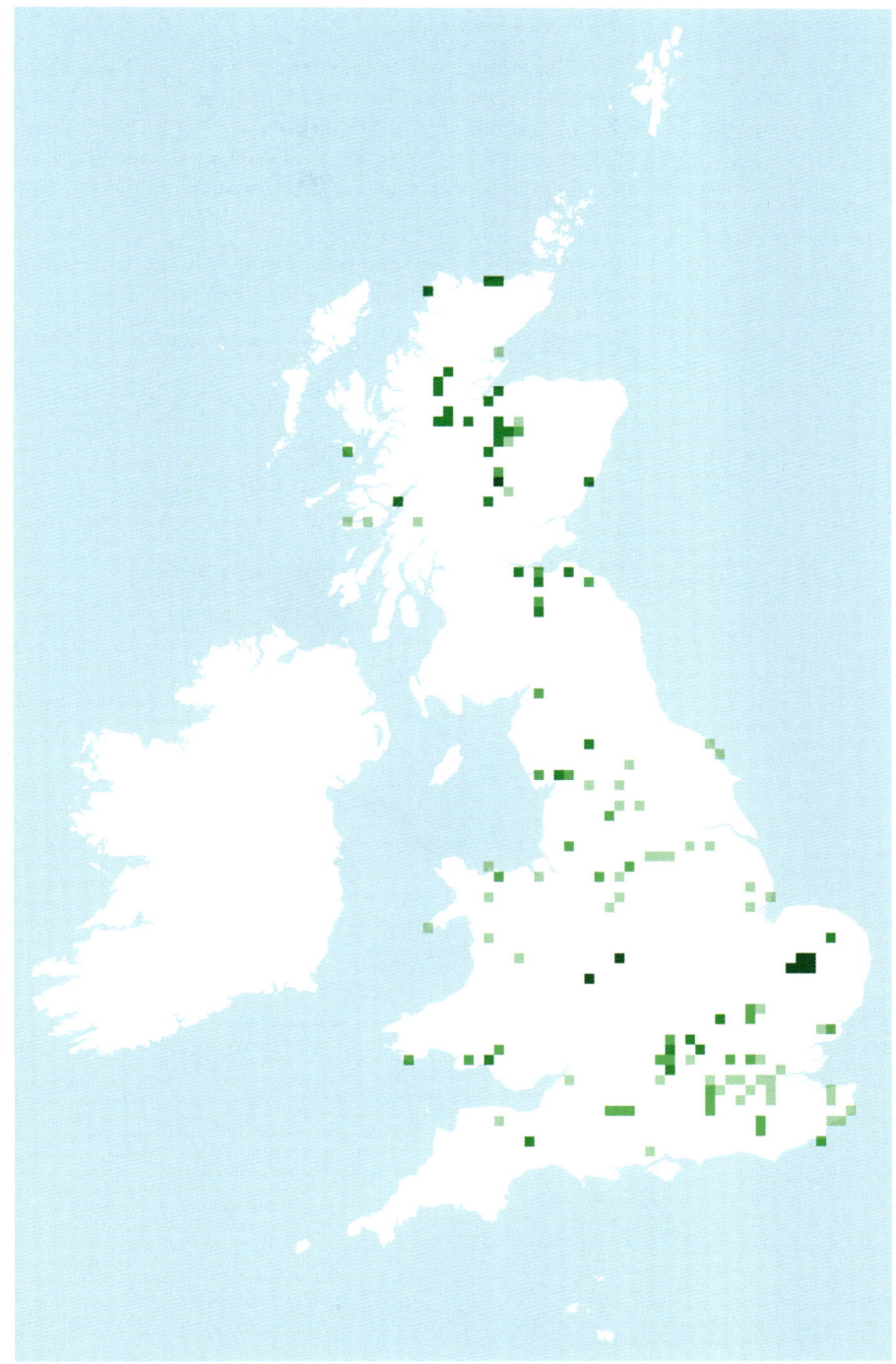

Map 36 *Myrmica lobicornis*

Myrmica lonae

There is some doubt as to whether this is a true species or a montane ecological form of *Myrmica sabuleti*. It has been found in Scotland and should be sought in further upland but sheltered locations. Little is directly known about its feeding habits, but they are probably essentially the same as its close relative *M. sabuleti*.

In the specimen depicted here, the middle leg on the left is missing, together with part of the right antenna.

Sizes: worker 4.0–5.0 mm; queen 5.5–6.5 mm; male 5.0–6.0 mm

Flight period: August–September

Map 37 *Myrmica lonae*

Myrmica rubra

One of the common red ants, with over 2,000 records on the NBN Atlas. A wide range of habitats is used, as long as it is quite warm and damp. It can become the most abundant ant in favourable locations. It feeds omnivorously, tending Homoptera and taking invertebrate prey. A very aggressive ant, especially where it is the dominant species, and it is likely to sting if the nest is disturbed. A wide range of myrmeco-philes has been found.

Sizes: worker 3.5–5.0 mm; queen 4.5–5.5 mm; male 4.5–5.5 mm

Flight period: July–August

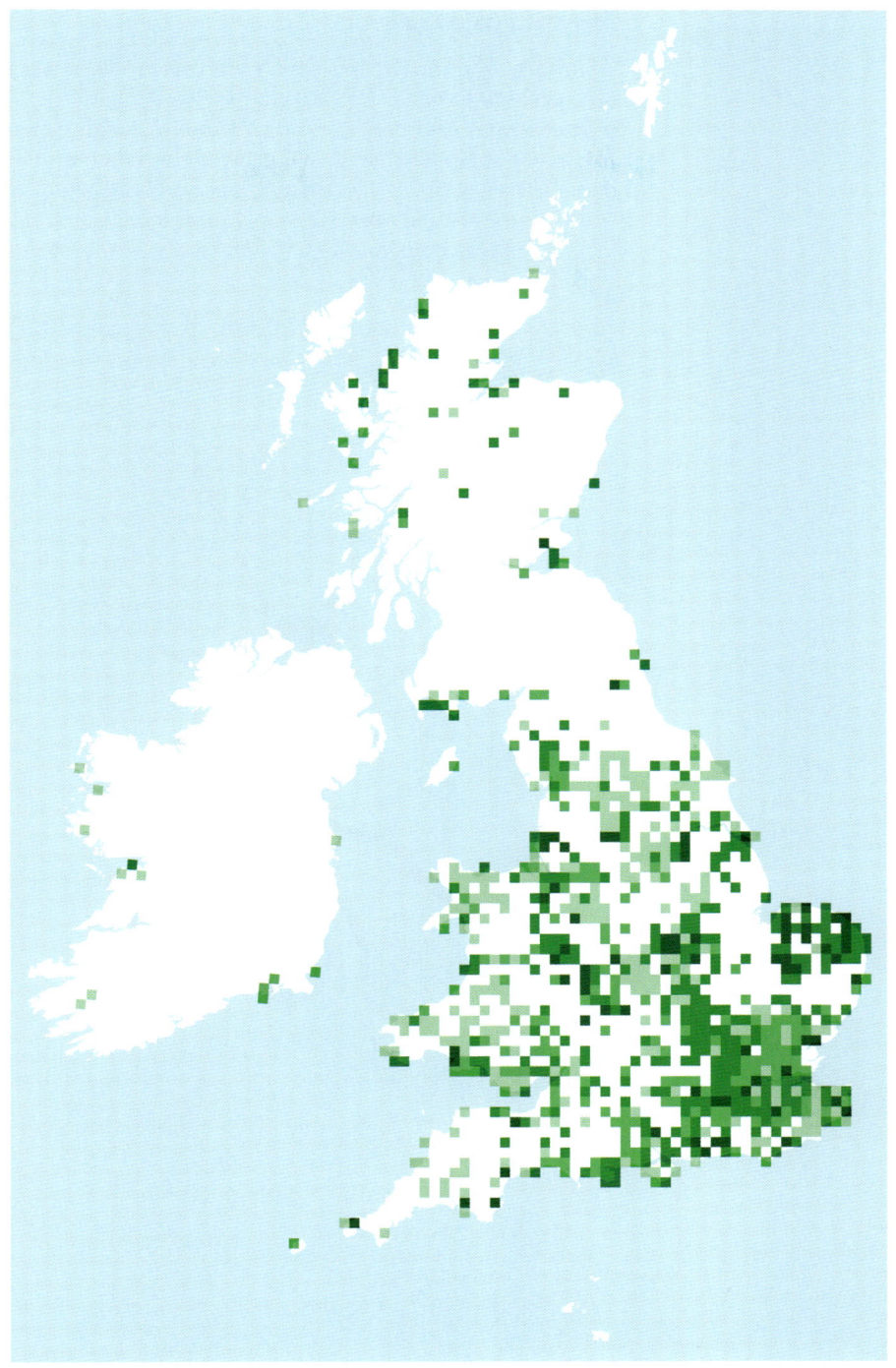

Map 38 *Myrmica rubra*

Myrmica ruginodis

One of the common red ants which, with over 6,000 records on the NBN Atlas, may be the commonest ant in the region. In line with its abundance, it can be found in a wide range of habitats, including being one of the few species found in damp, cool woodland. It is omnivorous and is also known to disperse seeds and take pollen, as well as tending aphids, including those living in trees. A wide range of myrmecophiles has been found.

Sizes: worker 4.0–6.0 mm; queen 5.0–6.0 mm; male 5.0–6.0 mm

Flight period: July–August

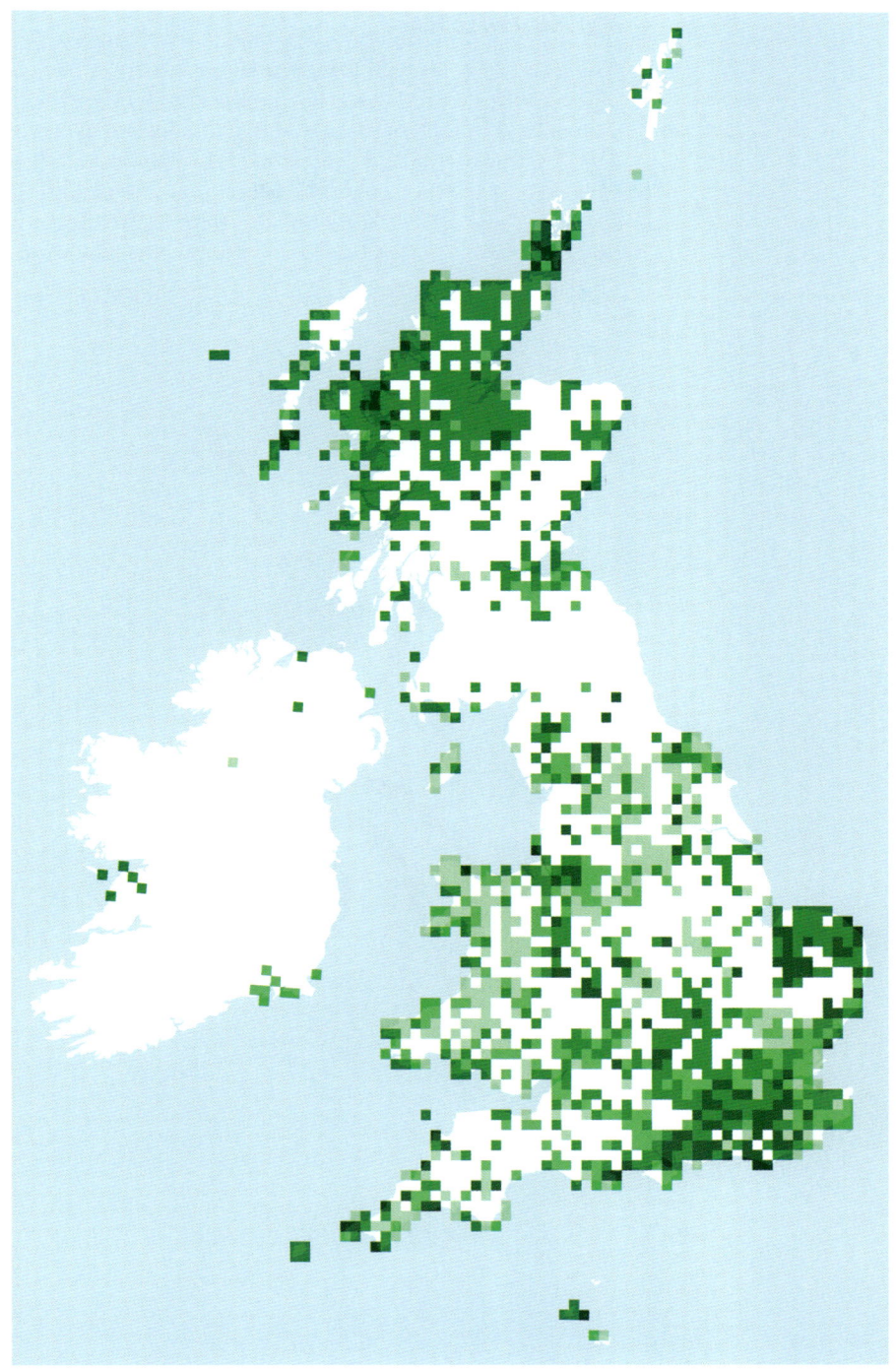

Map 39 *Myrmica ruginodis*

Myrmica sabuleti

Found in warm, dry habitats, this species tends to have a southern distribution but can be common throughout Britain and Ireland in suitably warm and sheltered locations. It feeds by scavenging and predation. This includes tending Homoptera, feeding at extrafloral nectaries and collecting seeds. It famously adopts the rare large blue butterfly *Phengaris* (formerly *Maculinea*) *arion* in its fourth larval stage.

Sizes: worker 4.0–5.0 mm; queen 5.5–6.5 mm; male 5.0–6.0 mm

Flight period: August–September

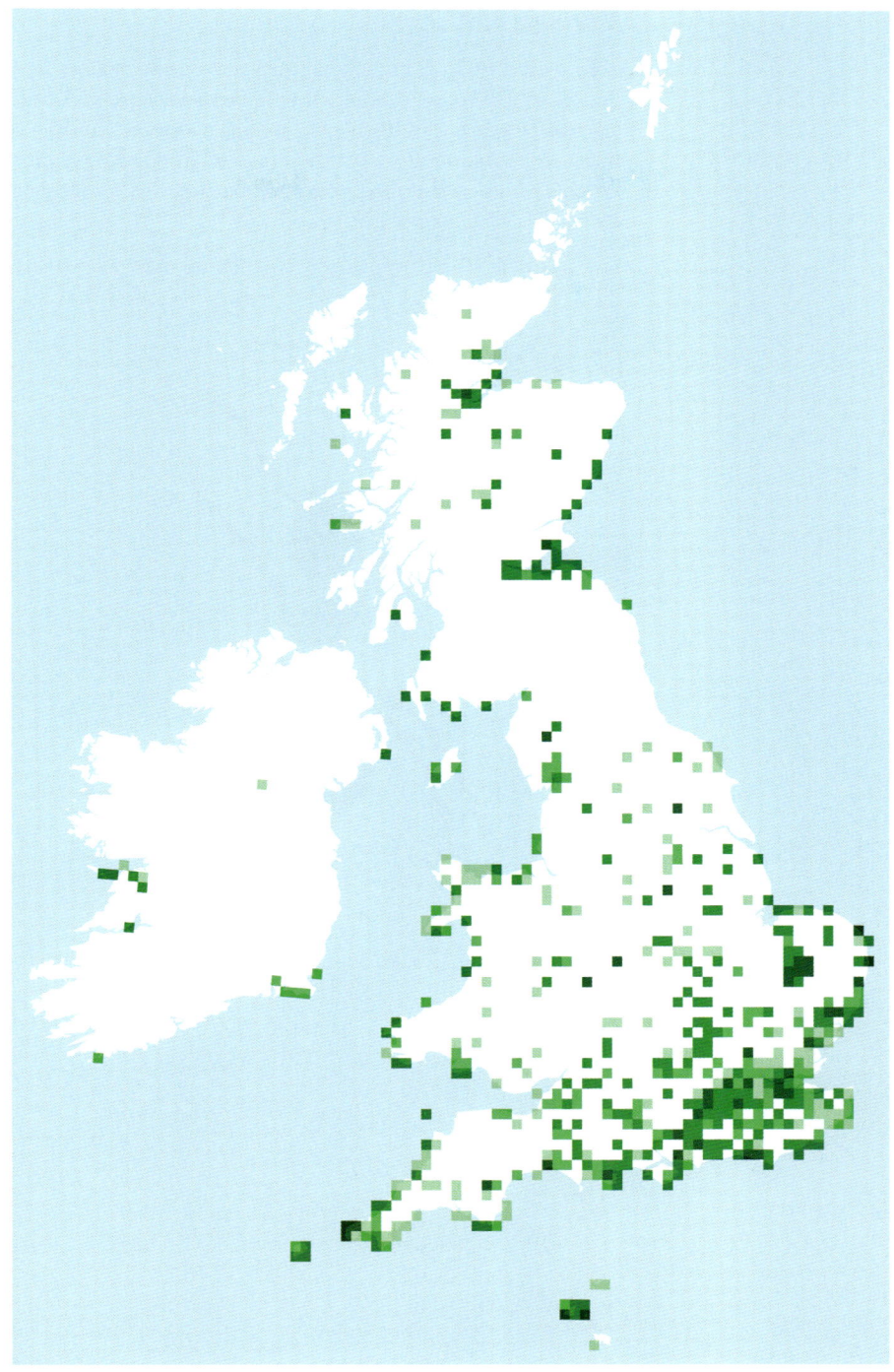

Map 40 *Myrmica sabuleti*

Myrmica scabrinodis

Found in a wide range of habitats across the region, but avoiding closed woodland, this species is dependent on aphids for honeydew and insect material as food. In bogs, it is known to frequently take insects caught by sundew plants. It has a range of myrmecophiles, including the woodlouse *Platyarthrus hoffmannseggii*.

Sizes: worker 4.0–5.0 mm; queen 5.5–6.5 mm; male 5.0–6.0 mm

Flight period: July–September

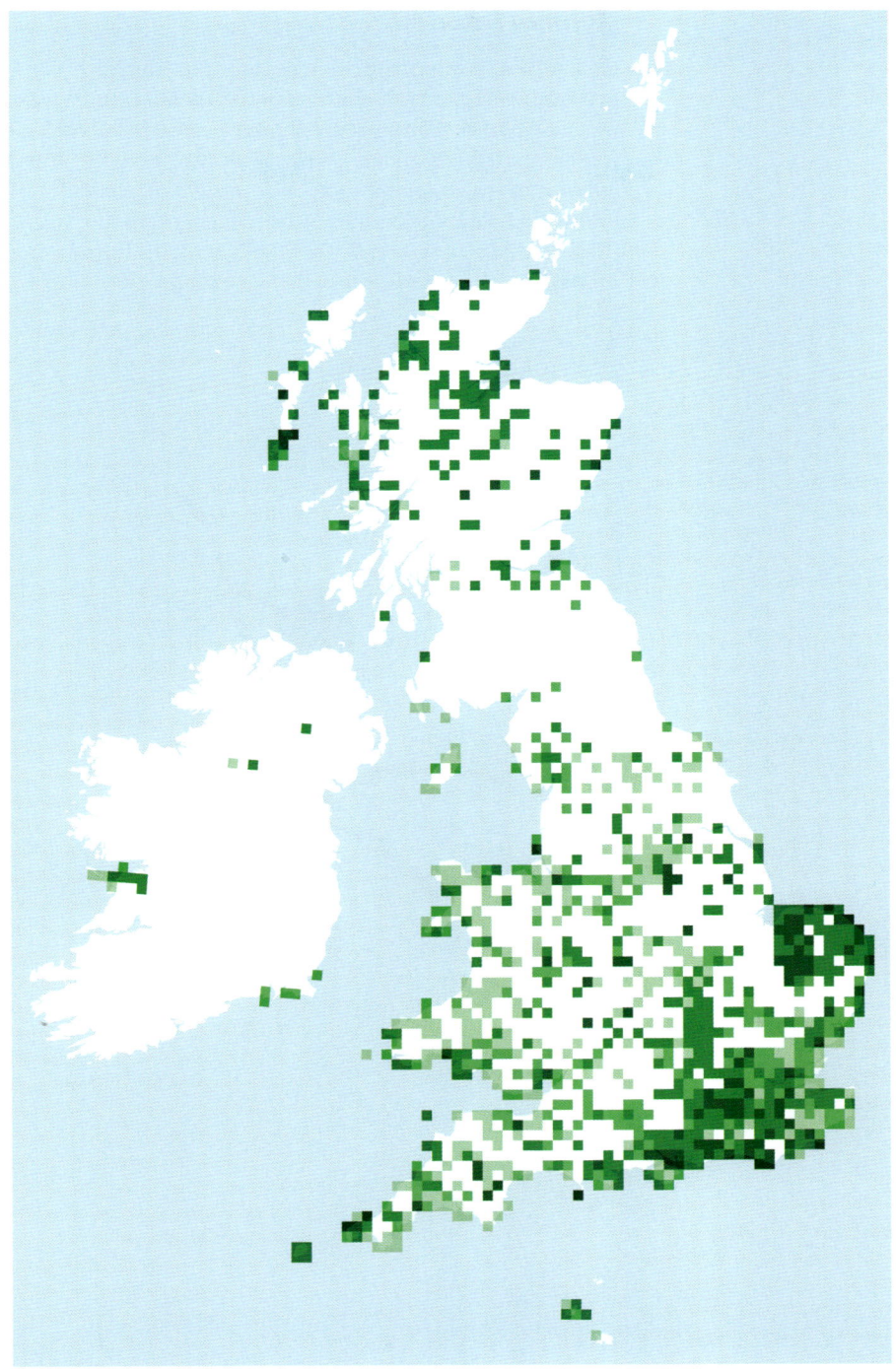

Map 41 *Myrmica scabrinodis*

Myrmica schencki

Generally a southern species in Britain and Ireland, this ant favours warm habitats with sparse vegetation. Its nest is unusual in having an entrance chimney constructed of plant fragments. It feeds mainly on invertebrates, including other ants, all of which are thought to be largely scavenged. It is known to host the mountain Alcon blue butterfly *Phengaris* (formerly *Maculinea*) *rebeli* on the mainland of Europe, but this species does not occur in Britain or Ireland.

Sizes: worker 4.0–5.5 mm; queen 5.0–6.0 mm; male 4.0–4.5 mm

Flight period: July–September

Map 42 *Myrmica schencki*

Myrmica specioides

A local species with records to date confined to warm sites, such as sand dunes, in coastal south-east England. Originally found only in east Kent, it has expanded its range significantly in recent years. Workers may be confused with those of *M. sabuleti*, but the males of these two species are easily separated. A very aggressive ant feeding mainly on other ants such as *L. flavus* and *L. alienus*. It will readily sting.

Sizes: worker 3.0–4.5 mm; queen 5.0–5.5 mm; male 5.0 mm

Flight period: August–September

Map 43 *Myrmica specioides*

Myrmica sulcinodis

A dark, strongly sculptured ant recorded widely throughout Britain but not seen so far in Ireland. It nests mainly in well-drained or gravelly soils in mountainous areas, but also on the edge of bogs in the south of England. Very little is known about its feeding, but it is assumed to rely mainly on small invertebrates, which it either scavenges or kills. Homoptera are uncommon in its habitats, and it has only rarely been seen tending them.

Sizes: worker 4.0–6.0 mm; queen 5.5–6.8 mm; male 5.0 mm

Flight period: August–September

Map 44 *Myrmica sulcinodis*

Myrmica vandeli

This is a very rare ant, which was added to the British list as recently as 2003. It has been found at only a couple of sites so far, one in England and one in Wales, which are characterised by being warm and boggy. There is no specific information on its diet, save that it is likely to be similar to that of its close relative *M. scabrinodis*.

Sizes: worker 4.0–5.0 mm; queen 5.5–6.5 mm; male 5.0–6.0 mm

Flight period: July–September

Map 45 *Myrmica vandeli*

Plagiolepis pallescens

Belonging to a genus that is common in southern Europe, the pygmy ant, *P. pallescens* is confined to the Channel Islands although there are reports of it or related species being found sporadically in southern England. It is found in dry areas and feeds on small arthropods and sugary solutions.

In the specimen depicted, the right rear leg is missing its terminal parts, others are folded under.

Sizes: worker 1.0–2.0 mm; queen 3.0–4.0 mm; male 1.5–2.0 mm

Flight period: June–July

Map 46 *Plagiolepis pallescens*

Ponera coarctata

The genus *Ponera*, represented by two species in the region, includes our most primitive ants. The slow ant, as it is sometimes known, forms small colonies in damp, warm habitats in the south of England, south-east Wales and the Channel Islands, being particularly characteristic of the clay soils of the Thames estuary in Kent and Essex. The food consists exclusively of invertebrate prey. Map 47 shows *P. coarctata* records that are either from a period after the separation of *P. testacea* or old records that have been re-examined and shown to be definitely *P. coarctata*. Map 65 (p. 207) shows older records that have not been re-examined plus newer records where the recorder has not been able to confidently separate the two species and therefore the ant recorded could have been either *P. coarctata* or *P. testacea*.

Sizes: worker 3.0–3.5 mm; queen 4.0–4.5 mm; male 3.4–3.8 mm

Flight period: August–September

Map 47 *Ponera coarctata*

Ponera testacea

Only fairly recently (2003) separated from *P. coarctata* as a full species, *P. testacea* is a paler ant found in drier habitats than its sibling species. Only 20 or so sites are currently known, mainly along the south coast of England. Little is known about the biology due to its recent separation. The map shows records that are definitely *P. testacea* as opposed to its sibling *P. coarctata* (see also Maps 47 and 65).

The front legs are folded under in this specimen.

Sizes: worker 3.0–3.5 mm; queen 4.0–4.5 mm; male 3.4–3.8 mm

Flight period: August–September

Map 48 *Ponera testacea*

Solenopsis fugax

The thief ant has been found in just a few very warm sites in the south of England but is commoner on the Channel Islands. It forms populous colonies. Although it is inconspicuous due to an entirely subterranean lifestyle, it is probably genuinely rare. It gains its main nutrition from the brood of the much larger ants in its area (hence its common name). However, it is also known to prey on other invertebrates and tend Homoptera. It makes very narrow tunnels into the nests of other larger species, along which the larger species cannot pass. In addition, when on a raid into a nest it produces a repellent, which effectively stops the prey species from defending the brood.

Sizes: worker 1.5–3.0 mm; queen 6.0–6.5 mm; male 4.0–4.8 mm

Flight period: September–October

Map 49 *Solenopsis fugax*

Stenamma debile

This is a predominantly woodland species. It is inconspicuous but most easily found by searching in or sieving leaf litter. The workers forage singly and scavenge, take prey and lick spilled honeydew. They are timid and act dead if disturbed. Prior to the separation of this species from *Stenamma westwoodii* in the mid-1990s, all British specimens were recorded under the latter name. There is just one Irish record. It is now known that virtually all older records are of *S. debile*, making *S. westwoodii* very rare. The map shows records that are definitely *S. debile* as opposed to its sibling *S. westwoodii* (see also Maps 51 and 66).

Sizes: worker 3.5–4.0 mm; queen 4.2–4.8 mm; male 3.8–4.2 mm

Flight period: September–October

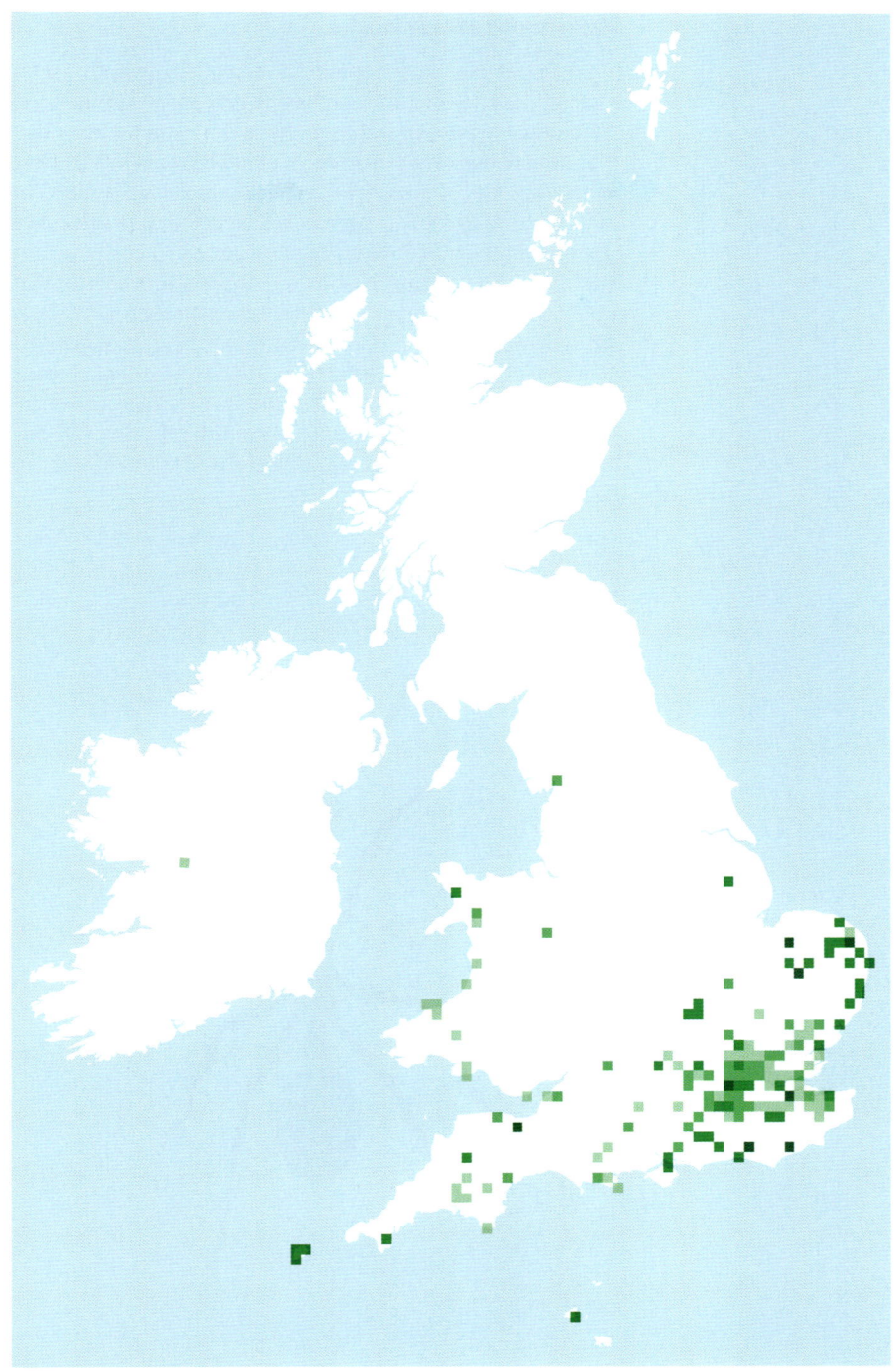

Map 50 *Stenamma debile*

Stenamma westwoodii

Now that it has been separated from the more common *S. debile*, *S. westwoodii* is thought to be very rare. Furthermore, it would now appear that England is the main home of this species, it being very rare in continental Europe. In addition, there is some evidence that the species may be in decline. In contrast to *S. debile*, this species is suspected to prefer more open habitats and so may not be detected so easily in leaf litter collections. Map 51 shows *S. westwoodii* records that are either from a period after the separation of *S. debile* or old records that have been re-examined and shown to be definitely *S. westwoodii*. Map 66 (p. 208) shows older records that have not been re-examined, plus newer records where the recorder has not been able to confidently separate the two species and therefore the ant recorded could have been either *S. westwoodii* or *S. debile*.

Sizes: worker 3.5–4.0 mm; queen 4.2–4.8 mm; male 3.8–4.2 mm

Flight period: September–October

Map 51 *Stenamma westwoodii*

Strongylognathus testaceus

A very rare parasite of *Tetramorium caespitum*, it is found only where there is a high density of the host nests. In such locations this species may occupy a significant percentage of host nests. The light-coloured workers contrast with the dark host workers. They are not produced in large numbers and do not forage. Other species of the genus (in continental Europe) are slave-makers, but this does not seem to be the case for *S. testaceus*. It is fed by workers of its host.

Sizes: worker 2.1–2.4 mm; queen 3.5–3.8 mm; male 3.2–4.0 mm

Flight period: July–August

Map 52 *Strongylognathus testaceus*

Strumigenys perplexa

This Australasian ant has recently been found living in outdoor locations on Guernsey. These tiny ants are found in leaf litter in this location and feed on small invertebrates such as springtails (Collembola). They are thought to have arrived on plants imported by a former local garden centre.

Sizes: worker 1.5–2.0 mm; queen 3.0 mm; male 3.0 mm

Flight period: September–October

Map 53 *Strumigenys perplexa*

Tapinoma erraticum

A rare (in Britain) ant confined to the south of England and the Channel Islands, *T. erraticum* looks like a small *Lasius* species but can be distinguished by its behaviour. Workers run around very quickly with the abdomen raised. It lives on open heaths where it feeds on carrion and prey, and occasionally tends homopterans. Map 54 shows *T. erraticum* records that are either from a period after the separation of *T. subboreale* or old records that have been re-examined and shown to be definitely *T. erraticum*. Map 67 (p. 209) shows older records that have not been re-examined, plus newer records where the recorder has not been able to confidently separate the two species and therefore the ant recorded could have been either *T. erraticum* or *T. subboreale*.

Sizes: worker 2.6–4.2 mm; queen 4.5–5.5 mm; male 3.5–5.0 mm

Flight period: June

Map 54 *Tapinoma erraticum*

Tapinoma subboreale

Only separated from *T. erraticum* in 2012, little is known of the biology of this species. It lives in similar habitats to its sibling species, although it can be found in gravelly areas not favoured by the other and is even more warmth-loving. Feeding is similar to that of *T. erraticum*. The map shows records that are definitely *T. subboreale* as opposed to its sibling *T. erraticum* (see also Maps 54 and 67).

In the specimen shown, the tarsus is missing from the middle leg on the left, and the legs on the right are folded under.

Sizes: worker 2.6–4.2 mm; queen 4.5–5.5 mm; male 3.5–5.0 mm

Flight period: June

Map 55 *Tapinoma subboreale*

Temnothorax albipennis

Older literature has this and the following three species in the genus *Leptothorax*. It is found almost exclusively on the coast and forms small colonies in a variety of cavities, including rock crevices and hollow stems of dead plants. It feeds mainly on scavenged and live invertebrates, although it has been found in abundance in flowerheads, so may take nectar. It has not been found in Ireland.

Sizes: worker 2.3–3.4 mm; queen 3.7–4.5 mm; male 2.5–3.2 mm

Flight period: July–August

Map 56 *Temnothorax albipennis*

Temnothorax interruptus

Markedly less common than its relative *T. albipennis*, *T. interruptus* is found in similar sites. It does not appear to nest in stems, the nest being hard to find in the soil. It feeds mainly on small invertebrates. Springtails (Collembola) are frequently stalked and jumped on.

Sizes: worker 2.3–3.4 mm; queen 3.7–4.2 mm; male 2.5–3.0 mm

Flight period: July–August

Map 57 *Temnothorax interruptus*

Temnothorax nylanderi

The most widespread of the four *Temnothorax* species, *T. nylanderi* used to be confined to central southern England but has undergone a dramatic spread in recent years. It is currently found scattered across England and Wales to the south midlands. It is generally encountered in shadier places than the other *Temnothorax* species and workers are often found in leaf litter collections. It nests in all kinds of hollow spaces, including dead wood, under bark of living trees, and in acorns. It is known to carry seeds of various plants, including various *Viola* species.

Sizes: worker 2.3–3.4 mm; queen 4.2–4.7 mm; male 3.0–3.2 mm

Flight period: July–August

Map 58 *Temnothorax nylanderi*

Temnothorax unifasciatus

An exclusively Channel Islands species. Two workers were found in Chiswick (west London), but this was probably an accidental introduction. It occurs in warm, open habitats and nests in all kinds of crevices. From here it forages for prey, drinks nectar and honeydew, and carries seeds back to the nest.

The funiculus of the left antenna and the middle right leg are missing in this specimen.

Sizes: worker 2.8–3.5 mm; queen 4.0–4.5 mm; male 2.8–3.5 mm

Flight period: July–August

Map 59 *Temnothorax unifasciatus*

Tetramorium atratulum

A workerless parasite which until very recently was named *Anergates atratulus*. It is found only where nests of its host (*Tetramorium caespitum*) are in high density, and then in only a few nests. The gynes and wingless males are fed by the host workers, to which they are similar in size. Newly mated gynes are thought to seek out queenless host colonies for invasion. Hence, *T. atratulum*-infested colonies only last as long as the host workers survive.

This specimen is a queen, and missing its left antenna and the right hind leg.

Sizes: there are no workers; queen 2.5 mm; male 2.3 mm

Flight period: June–July

Map 60 *Tetramorium atratulum*

Tetramorium caespitum

A mainly coastal and heathland species, this ant forms populous colonies and is often dominant where found. In Scotland, *T. caespitum* is recorded from south-facing igneous rock outcrops on or near the coast. Although workers of this northern population are visually indistinguishable from those of southern populations, new evidence based on examining males suggests it comprises a distinct species of *Tetramorium* new to Britain. At Dungeness, Kent, in 2021, of 30 roof tiles laid on the ground, 18 had *T. caespitum* nests. The species is aggressive and feeds as a predator, scavenger, and a tender of aphids, as well as collecting seeds. A closely related species, *Tetramorium impurum*, was reported in 2019 from Guernsey (Attewell & Wagner 2019). Workers appear almost identical to *T. caespitum*.

Sizes: worker 2.5–4.0 mm; queen 6.0–8.0 mm; male 5.5–7.0 mm

Flight period: June–October

Map 61 *Tetramorium caespitum*

Map 62 Records that could be *Hypoponera punctatissima* or *Hypoponera ergatandria*

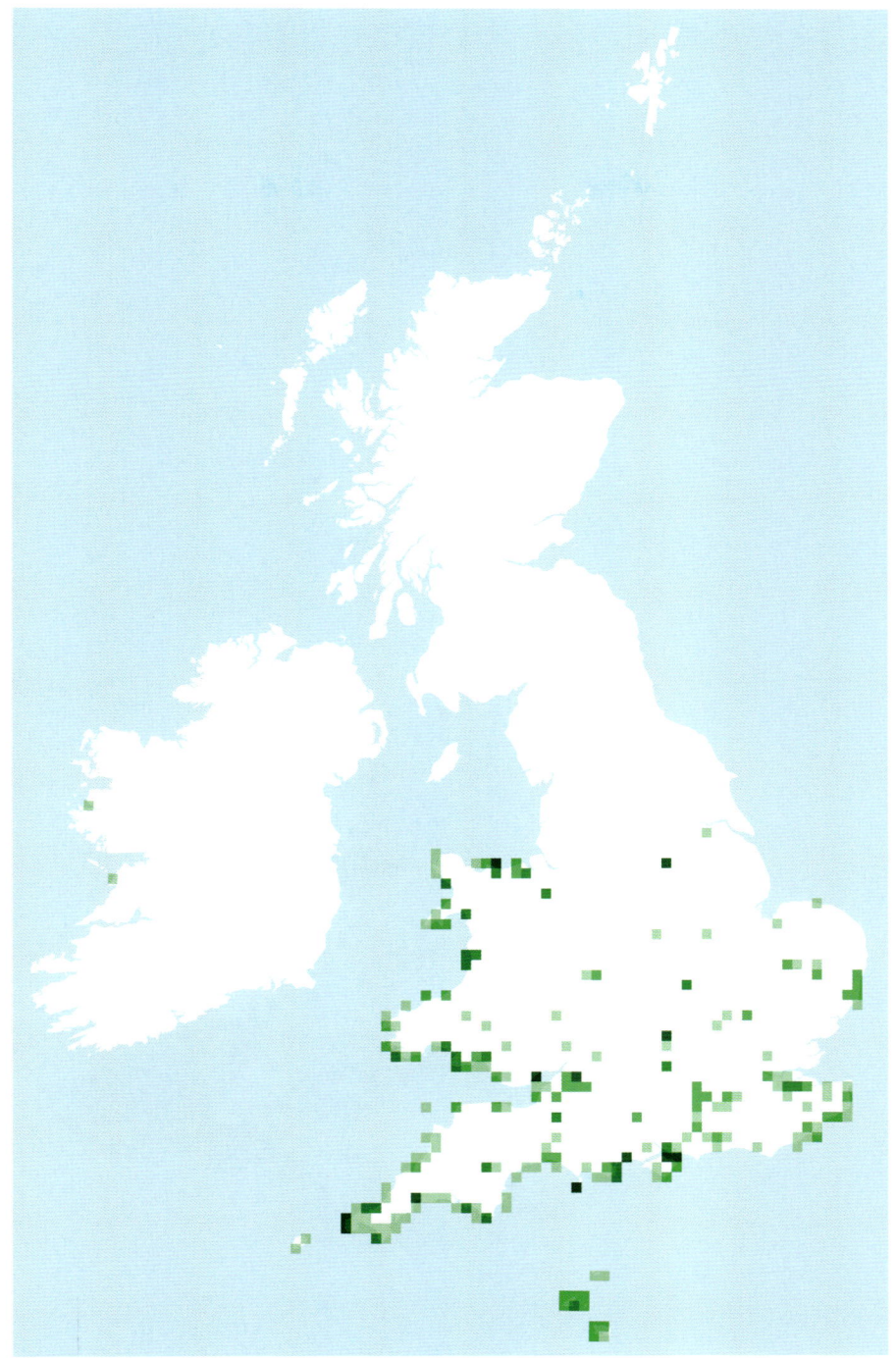

Map 63 Records that could be *Lasius alienus* or *Lasius psammophilus*

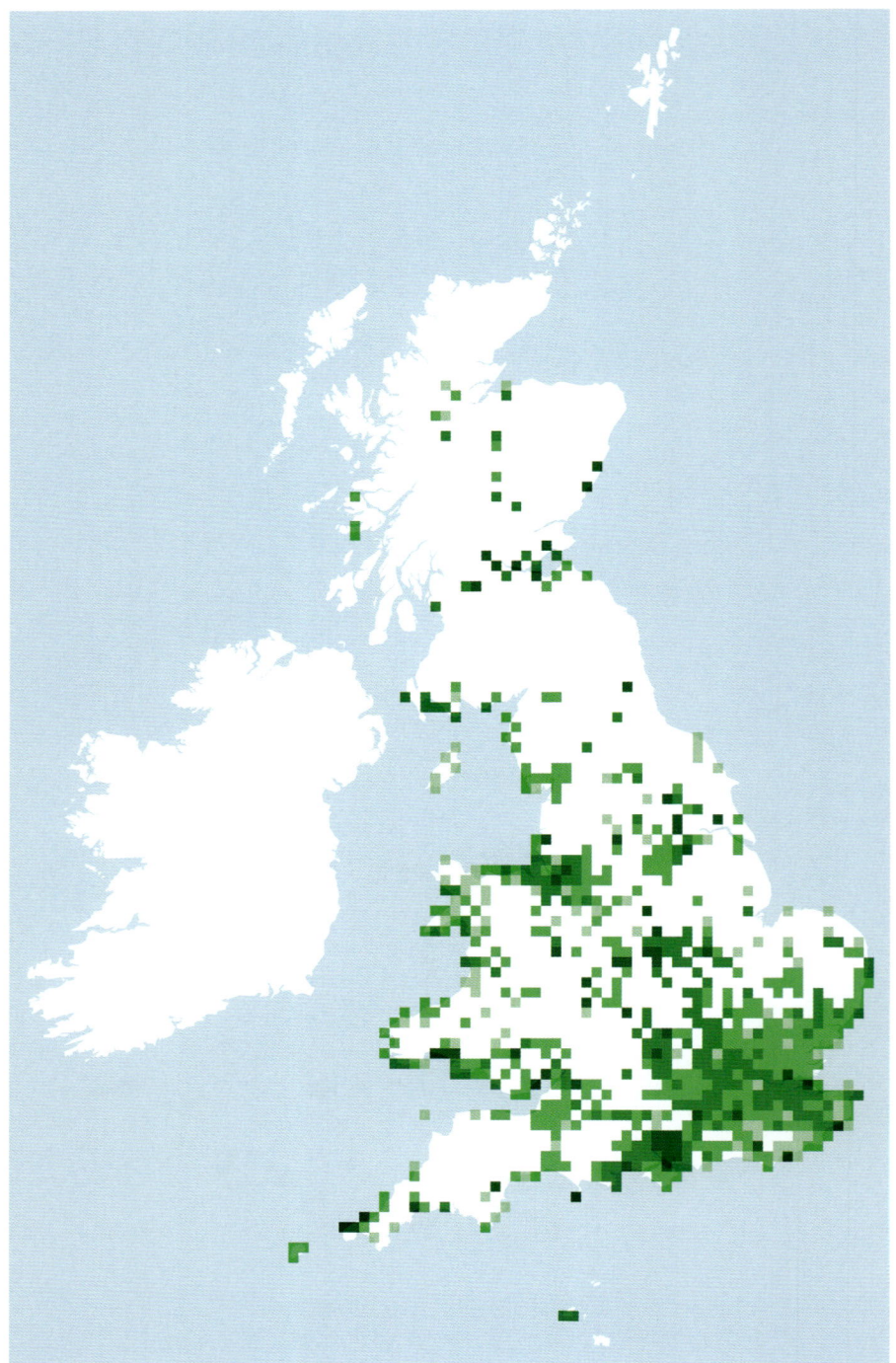

Map 64 Records that could be *Lasius niger* or *Lasius platythorax*

Map 65 Records that could be *Ponera coarctata* or *Ponera testacea*

Map 66 Records that could be *Stenamma westwoodii* or *Stenamma debile*

Map 67 Records that could be *Tapinoma erraticum* or *Tapinoma subboreale*

4.2 Tramp species

Commerce and international travel have conspired to spread many animals and plants around the world, often many thousands of miles from their true home. When these are ants, they are referred to as tramp species. Those that have been found outside in Britain and Ireland are included in the key in Chapter 6. Those that have been encountered only in various kinds of hothouse are discussed briefly here. Unsurprisingly, the list of what should be regarded as a tramp in Britain and Ireland constantly changes and is variable. In their 1975 key to British ants, Bolton & Collingwood (1975) named nine tramp species, together with some unnamed species in the genus *Camponotus*:

- Some *Camponotus* species
- *Crematogaster scutellaris*
- *Linepithema humile* (formerly *Iridomyrmex humilis*)
- *Monomorium pharaonis*
- *Nylanderia vividula* (formerly *Paratrechina vividula*)
- *Paratrechina longicornis*
- *Pheidole megacephala*
- *Tapinoma melanocephalum*
- *Tetramorium guineense*
- *Tetramorium simillimum*

Brangham (1938) listed 38 such species at Kew Gardens, but many of these were just occasional specimens found only once or twice. The suggestion from an expert on the tramps (Matt Hamer, personal communication 2021) is that the following are the key species to highlight as having been seen in Britain:

- *Crematogaster scutellaris*
- *Monomorium floricola*
- *Monomorium pharaonis* (on the BWARS list and dealt with above)
- *Pheidole megacephala*
- *Pheidole pallidula*
- *Plagiolepis alluaudi*
- *Technomyrmex difficilis*

Some notable tramp species are described below.

At the Eden Project in Cornwall there is hardly anywhere where the white-footed ant (*Technomyrmex difficilis*) is not seen (Fig. 4.4). The species is found nowhere else in Britain or Ireland at the time of writing. Originating from Madagascar, this ant has become widespread outdoors around the world but will be unable to survive outside in Britain or Ireland, at least for the time being.

Fig. 4.4 White-footed ant, *Technomyrmex difficilis*, common in Eden Project hothouses

Although there are no occurrences reported by modern sources, the highly invasive species *Pheidole megacephala* was listed by Bolton & Collingwood (1975) and it is likely that it still occurs sporadically, but only so far in heated situations. Lebas *et al.* (2016) report that it occurs outside on the Mediterranean coast and throughout the region in heated buildings. This is a highly invasive species, so let us hope that climate change does not make a suitable environment for it outside in Britain or Ireland. Perkins (1913) noted that in Hawaii, where the species had been known since the 1870s, *'It may be said that no native Hawaiian Coleoptera insect can resist this predator, and it is practically useless to attempt to collect [beetles] where it is well established. Just on the limits of its range, one may occasionally meet with a few native beetles, often with these ants attached to their legs and bodies, but sooner or later they are quite exterminated from these localities.'* Like the recently invading *Lasius neglectus* (p. 128), *P. megacephala* can form supercolonies. It originates from Africa.

The Mediterranean acrobat ant, *Crematogaster scutellaris*, occurs as far north as central France (Lebas *et al.* 2016), and it has been seen in Britain in recent times. Two colonies were found in 2018 in the West Midlands and in London, and a queen was found, albeit in a package, in Birmingham in 2020. Collingwood (1979) states that it is often found in imports of cork.

Alluaud's little yellow ant, *Plagiolepis alluaudi*, has been found on a number of occasions in heated situations, including zoos and glasshouses. For example, in 2015 it was found at a butterfly farm in Stratford-upon-Avon, Warwickshire, and then, in 2016, at London Zoo. In these situations, it was regarded as 'a nuisance' and controlled. Like the white-footed ant, *Technomyrmex difficilis*, it is native to Madagascar.

The subject of tramp species is huge and rapidly changing; this account has just picked out a few examples. Wong *et al.* (2023) state that at least 520 species of ant have been moved out of their native range and that 60% of these become naturalised. In addition, Britain seems to be a hotspot for arrival of such species, although this may be in part due to the high vigilance of the border authorities and others.

4.3 Abundance

Colony density is very variable. For instance, *Tetramorium caespitum* has densities as high as 1 nest per m². *Lasius flavus* is thought to achieve a biomass higher than any other ant species in the world. Its nest densities are like those of *Tetramorium* species, but higher numbers of workers in nests mean that the biomass of *L. flavus* can be as high as 165 kg fresh weight per hectare. These ants can move as much as 7 tonnes of soil to the surface per hectare per year.

Methods for the estimation of colony number and colony size are described in Chapter 7. Application of some of these methods to size have shown that *Myrmica* colonies average about 1,000 workers. *Tetramorium* species, on the other hand, have about 10,000 workers on average, with some colonies studied having more than 30,000. *Lasius flavus* and *L. niger* have an average of about 20,000 and 14,000 respectively, with maximums of 100,000 and 60,000. Most of our other species have just a few hundred workers. But the champions of nest size in Britain and Ireland are the wood ants. *F. rufa* is quoted by Seifert (2018) as having up to 120,000 workers in a monogynous nest. It must be remembered that the English species *F. rufa* is almost certainly a *F. rufa* × *F. polyctena* hybrid (p. 36). Nests of continental *F. polyctena* have been shown to have over 17 million workers! It is likely that the hybrid will have something between the numbers of *F. rufa* and those of *F. polyctena*. A nest of *Formica pratensis* in Germany was estimated to have over three million workers. If this sort of size were mirrored in the Channel Islands populations of this species, it would hold the record for colony size in the British Isles. At the other end of the scale, nests of *Ponera* species have fewer than 100 workers.

When we consider gynes, numbers per nest may be very high in species that have low numbers of workers. At the extreme of *Tetramorium atratulum*, which has no workers, up to 3,000 gynes have been recorded. The *F. polyctena* nest with 17 million workers was estimated to have at least 20,000 gynes.

5 Conservation and management

Most species of any group of living things are under threat in modern, post-industrial Britain and Ireland. Habitat loss and poor management, pollution and climate change ensure that this is the case. Ants are no exception.

Although ants appear to be everywhere, some of our species are in serious trouble. One way of looking at this is to learn that nine UK species had Biodiversity Action Plans (BAPs), now covered under the UK Post-2010 Biodiversity Framework (Benyon *et al.* 2012) (Table 5.1). BAPs are no more as such, but these ants are still in the same position as they were. UK BAP priority species were those that were identified as being the most threatened and requiring conservation action. In Ireland, similar concerns are felt.

This represents nearly 15% of the regional list. In addition to this, many other species, especially some of the other wood ants and their relatives, are clearly suffering a decline. The account that follows looks at the nine BAP species, the two non-BAP wood ants and then some other threatened species designated in Shirt (1987) and Falk (1991).

In the following pages, the status of selected species will be considered. However, there is a wide range of factors that may affect both these and other ant species in Britain and Ireland. There is no question that loss of habitat is high on the list of reasons for the contraction of the range and distribution of ants in general. One relatively obvious situation is when the habitat that provides the bulk of the ants' food is lost or degraded. For example, in the wood ants, which are predominantly forest-dwellers, the loss of these forests immediately leads to the loss of much of their food in the form of honeydew. Poor management leading to overgrowth in woodland is also a threat to wood-ant survival. Grubbins Wood in south Cumbria, for instance, is thought to be now too shady for the wood ants for which it was once famous. A detailed analysis of the consequences of the loss of suitable habitat for wood ants can be found in Sorvari (2017).

Another important habitat for ants is heathland, particularly in the south on sandy soils. These warm habitats are ideal for many of our species. Such heathland, common in the past, has in most places been systematically destroyed due to its unsuitability for any kind of agriculture and its relative suitability for housing estates and other developments.

Table 5.1 UK Biodiversity Action Plan ant species

Formica aquilonia
Formica exsecta
Formica picea
Formica pratensis
Formica rufibarbis
Formicoxenus nitidulus
Tapinoma erraticum
Temnothorax interruptus
Tetramorium atratulum

Specific examples of the consequences of this loss can be found in the species accounts below.

A further threat to native ants is the legions of non-native or tramp species that are now found (p. 210). A recent example will suffice to show the problem. Looking much like the very common *Lasius niger*, the invasive *L. neglectus* may prove to be a coming plague in future years (p. 128).

Climate change, a factor affecting distribution and abundance in its own right, must also be considered. In this respect, ants may be one of the few beneficiaries, although as always in nature, the situation is complex and notoriously difficult to predict. In a recent report, Pearce-Higgins *et al.* (2015) concluded that in excess of 70% of British and Irish ant species were likely to benefit from future climate change. However, Parr & Bishop (2022) paint a much less optimistic picture for ants worldwide. In any case, ecosystems are likely to change in complex and unpredictable ways, even if some species seem to show an initial benefit.

Another aspect of climate change, especially warming, is that ants from the south may be more easily able to invade northwards. Jones (2022) has speculated about the possibility of a number of species that do not quite make it to Britain or Ireland presently, being able to do so as the climate warms. He includes in his list *Aphaenogaster subterranea* and *Messor capitatus* (both already on the Channel Islands), various *Temnothorax* species including *Temnothorax parvulus*, and *Camponotus ligniperda*.

5.1 UK BAP Species

Tapinoma erraticum and *T. subboreale*

The latter of these species was recently separated from *T. erraticum* by Seifert (2012). Both of them are always going to be restricted to lowland heath, so their continued survival is very much dependent on the conservation of this rather specialised habitat. People who live in or visit these areas could well contribute here, especially in relation to any differences in biology between *T. erraticum* and *T. subboreale*, about which little is known. This could be crucial information, as the first step in any conservation effort is to gain detailed knowledge of the species and where it lives.

Tetramorium atratulum

This ant is found only in or around the nests of the species it parasitises, *Tetramorium caespitum*. This restricts it to the range of that species, which is generally southern (in

England and Ireland) and coastal, including several sites in the Firth of Forth area of Scotland (see *Tetramorium caespitum* map on p. 203). However, if this map is compared with the *T. atratulum* map on p. 201 it becomes clear that *T. atratulum* is much more restricted in range than its host. Falk (1991) suggests that it is likely that the parasite requires very dense populations of its host. These are found in only parts of southern England. Brian (1977) observed that even in France, where the host is much more common, few nests harbour *T. atratulum*. Nonetheless, the ant is also probably somewhat under-recorded as a result of its secretive habits and the difficulty of finding it because of the need to excavate host nests. There is little doubt that the main habitat – lowland heath – of the host and thus the parasite is under threat. Careful management of this habitat is thus crucial for the continued survival of this very rare ant.

Temnothorax interruptus
Another very rare ant, this species is neither easy to find nor easy to study. A 2018 report (Munns *et al.* 2018), which looked at 10 sites where it was known in the past, reported only three sites with nests and a total of just six nests. This is an important study as it looks at one of the few sites where *T. interruptus* is currently known to occur. It suggests that the uncommon habitat of dry open heath with short vegetation is under threat from a variety of factors including succession, fires, increase in nutrients and changes in grazing and trampling. Thus, as with other species, this ant's continued survival depends on conserving the habitat in which it lives. Recommendations of the report are for further monitoring, consideration of the possibility of translocations, maintenance of grazing regimes and the control of succession, the latter tending to shade the ants out.

Formicoxenus nitidulus
As discussed (p. 106), this species is always associated with the red wood ants. Naturally, therefore, it will benefit from the measures employed to conserve its hosts. There is little doubt that this species is under-recorded due to its small size and its nesting habit deep inside the nest of a much larger and more numerous species. It would be worthwhile looking for this species anywhere where true wood ants are found. Choose a warm but somewhat overcast day from August to October. Very careful observation of the nest surface should reveal the tiny workers and males running about. The only possible confusion would be with very small wood ant

workers, but they are neither shiny nor uniform in colour, and never as small as *F. nitidulus* (see Fig. 3.16).

Formica exsecta

At the time of its inclusion in the UK BAPs in 1995 (Plowman 1995), this species was known from a few sites in Devon and pine forests in Strathspey (Figs 5.1, 5.2). This itself was a decline from earlier times when it was common in the New Forest area and in Cornwall. By 2008, Chudleigh Knighton Heath SSSI in Devon was the only place left in England where *F. exsecta* could be found (Carroll 2008).

Since then, a lot of research and conservation has been done. The result of this is that two new sites for the ant, both in Devon, have been established. To achieve this success, much work on habitat suitability and ways to propagate the ant had to be undertaken. This species has quite a complex reproductive biology and, before these studies, the details were little known. It now appears that many nests do not produce queens. In 2020, from 200 nest structures in Devon, only 17 produced queens and only nine produced males. Queens and males emerged in the morning throughout just

SSSI
Site of Special
Scientific Interest

Fig. 5.1 A nest of *F. exsecta* in typical habitat in the Spey Valley area, Scotland

Fig. 5.2 Worker of *F. exsecta* showing distinctive notched occiput, the back of the head

a week or so in July. Such information helps with attempts to breed the ants. A variety of approaches were used to try to increase the range of this rare species, although most had major drawbacks (Carroll 2008). These methods included:

- supplementary feeding of nests with an egg and honey mixture;
- use of artificial solaria by placing plastic covers over nests;
- release of mated queens;
- creating artificial nest sites for the host ant species, *F. fusca*;
- provision of nest material such as shredded dried grass;
- translocation of nests;
- heathland management.

Buglife have continued along these lines and have produced two more detailed reports (Vulliamy 2020; Walters 2020).

In England then, the situation is that the last natural population is secure and two new sites have been colonised, but it is too early to say if either of these will be self-sustaining (Carroll 2021, personal communication).

In Scotland, the outlook is somewhat less precarious, but there is no reason for complacency. The high number of records on the NBN Atlas is likely to be somewhat inflated by the intense activity of certain groups in the relatively small area where the ants occur. For example, the Highland Biological Recording Group (HBRG) contributed 226 records between 2000 and 2022. The RSPB reserve at Abernethy has contributed a further 320. These big numbers rather obscure the relative rarity of this species, even in Scotland, where it warrants a place on the Scottish Biodiversity List (SBL) (NatureScot 2012). Much research has also been done on the Scottish populations, although it is largely to be found in unpublished reports and theses. These can be found listed at https://www.woodants.org.uk/literature.

In an overview written in 2021, Wiswell (personal communication) updated the situation at the five remaining localities surveyed by Hughes (2006). Hughes had previously failed to find the ant at two other sites in Scotland, which had harboured it in the 1950s (Hughes 1997). By 2021, Wiswell reported that numbers at two of these five sites were 'declining', probably due to 'habitat succession shading nests' and flooding. *F. exsecta* needs temperatures of up to 30°C for brood development and cannot achieve this by the production of metabolic heat, unlike the wood ants, and so needs an open habitat (Siefert 2000). One other site was described as 'at risk due to habitat succession'. Only one of the five sites was regarded as 'stable'. Clearly in Scotland, too, *F. exsecta* needs to be looked after. Possible measures would include increased grazing and scrub removal to slow habitat succession.

A detailed list of the scattered *F. exsecta* literature can be seen at http://hymettus.org.uk/Formica_exsecta.htm

Formica picea

The black bog ant looks very like the common *Formica fusca* and *Formica lemani* but individuals are much more shiny and less hairy than these two species (Fig. 5.3). In the field, it cannot be mistaken due to its very unusual nests. These are topped with small domes of dead vegetation, built around living vegetation, in boggy areas (see Fig. 3.6). Donisthorpe (1915) lists it as being present on the Isle of Wight, at Wareham in Dorset, around Bournemouth and in the New Forest, together with a site in Wales at Rhossili, on the western end of the Gower. By the early 1990s this very rare ant was described as being 'found in Dorset with one record in Wales' (Skinner & Allen 1996). The record in Wales was reported in 1913 and had never been confirmed and was doubted by experts in later years. Then, in 1991, the ant was spotted on

Fig. 5.3 A worker of *Formica picea*

a raised mire at Cors Goch Llanllwch in Carmarthenshire. Subsequently, a thriving population has also been found at the original 1913 site of Rhossili. The Cors Goch Llanllwch site is probably the largest colony in Britain and appears to be healthy. In June of 2021, over 50 active nests were counted in one small area of the mire in the space of about two hours (Fig. 5.4).

Conservation efforts for this species are generally still at the monitoring and study stage (e.g. Fowles & Hurford 1996 and Rees 2006), although some drainage-channel blockage has been carried out at Cors Goch Llanllwch to maintain the water table which, it appears, is an important ecological factor for this species. Recommendations have been made for the translocation of the species into new areas after detailed ecological assessment of their suitability. No reports of such translocations could be found during the research for this book. However, there was good news in 2003 when the ant was found in Surrey. This species is probably safe from extinction as long as its habitat remains protected, but it is unlikely to increase in numbers without intervention, such as nest translocation.

Fig. 5.4 Part of the Cors Goch Llanllwch National Nature Reserve and SSSI where *F. picea* nests are found in abundance

Formica rufibarbis

This species, which was once described as being 'possibly the rarest resident animal in mainland Britain' (Pontin 2005), was until recently thought to have become extinct there. Although it was never common, Donisthorpe (1915) recorded it from Reigate, Ripley, Chobham and Weybridge. These sites are all in Surrey and are consistent with the modern distribution, which is only in Surrey. Yarrow (1954) also restricts it to this area, together with a population on the Isles of Scilly (on the island of St Martin's, Fig. 5.5). Pontin (2005) states that it was present at Oxshott in 1965 and at Chobham in 1967, this latter nest having been destroyed by a rubbish dump by 1968. Finally, it was refound on Chobham Common and at Stickledown rifle range in 1992. In 2005 these two sites still had the ant. Yet by 2017 it was lost from both.

Fig. 5.5 Chapel Down on St Martin's, Isles of Scilly, a stronghold of *F. rufibarbis*

The final piece of (good) news is that in 2017, just when the last ants were seen in Surrey, a new site in Hampshire was discovered (Dodd & Kirk 2017, personal communication). This makes *F. rufibarbis* again possibly the rarest animal on the mainland!

In the meantime, extensive conservation efforts seem to have come to nought. There was optimism when Gammans reported on progress to date in 2008 (Gammans 2008). However, Dodd (2015) reports that 40 nests of *F. rufibarbis* were released into a prepared site at Chobham Common between 2007 and 2010. Sadly, the project failed and by 2014 no colonies had established.

The reasons for the red-barbed ant's decline in Britain include loss of suitable heathland habitat through human development, as well disturbance and inbreeding caused by the small size of the ants' populations.

Formica pratensis

This is another species that was always rare in Britain, being only ever found in the Bournemouth area. It became extinct there in 1987 and now occurs on only a few clifftop sites in the Channel Islands. Here a decline has been noted since the 1990s. In a survey on Guernsey in 1990 nearly 100 nests were seen, but by 2017 this was down to fewer than 60 – a 42% decline. In 2018 an action plan was put in motion. Measures included the marking of each nest with red flags (Fig. 5.6). These are to try to ensure the area does not get strimmed when the paths are being cut back, and to allow the careful trimming of the vegetation that threatens to overwhelm the nest in spring. In 2021, 76 nests were seen, so the early signs were that the measures may be having a positive effect.

In the Biodiversity Action Plan for this species, it was suggested that a captive breeding programme followed by reintroductions may be initiated. No such programme seems to have ever been carried out and the species remains extinct on the mainland.

Formica aquilonia

Of the four members of the red wood ant group, only *Formica pratensis* and *Formica aquilonia* had Biodiversity Action Plans. In the case of *F. aquilonia* the situation is more encouraging. The species appears quite healthy in its stronghold in the Highlands of Scotland (Fig. 5.7). The main threats to its existence are the loss of its pine woodland habitat and poor management. Maintenance and sympathetic management of that habitat are all that is fundamentally needed to ensure

Fig. 5.6 *Formica pratensis* nest on a clifftop path, south Guernsey. The red flags are to try to ensure that the area does not get strimmed when the paths are being cut back and during the trimming of the vegetation, which threatens to overwhelm the nest in spring.

Fig. 5.7 A huge nest of *Formica aquilonia* in pine woodland, Spey Valley, Scotland. The fern trowel is 25 cm long.

its continued survival. Hopefully, in this more conservation-minded time, this should be achievable.

There is just one site in Ireland, in Annaghgarriff Wood, part of Peatlands Park in north Armagh, which harbours about 40 nests of *F. aquilonia*.

5.2 Non-BAP Red Wood Ants

All four British species of red wood ants seem to be under pressure. Two of them had Biodiversity Action Plans, as we have seen above. The other two are dealt with below.

Formica rufa

Although not the subject of a BAP, there is no question that this species is struggling and has been lost from some sites, especially in the north of England. The most obvious of these are its northerly extensions in the Lake District. None of the former sites in this region have *F. rufa* anymore (see the *F. rufa* map on p. 101). There is a suggestion, which seems to be well supported, that these sites were in fact never the natural habitat of *F. rufa* but that the ants were there because of the introduction of this species in Victorian times for the purposes of rearing pheasants, a very common practice then.

A little further south, however, where there is no such suggestion, the species also seems to be in decline. A population that was studied in the Silverdale, Lancashire, area in the mid-1970s and early 1980s has now completely disappeared. Just north of this area, in Gaitbarrows Wood, Lancashire, *F. rufa* is fortunately still found.

In a wider-ranging survey, covering the midlands and north of England, Robinson (2011) visited the sites of 30 historical records and found only 18 still with nests (a 40% decline). The situation further south in the UK seems to be less worrying, but a combination of habitat loss and inappropriate management could always put this keystone species under threat, so we must not be complacent.

In the far south of England, *F. rufa* is doing well, with strong populations being found across the southern counties and mid-Wales (Fig. 5.8). Moreover, some successful translocations have been carried out with *F. rufa*. Attewell (2019, 2020) reports on two such projects in Hertfordshire. The experience gained from these and other projects is summarised by Wiswell *et al.* (2022).

Formica lugubris

This species has its strongholds in the north and midlands of England, and in the highlands of Scotland (Fig. 5.9).

Fig. 5.8 A *Formica rufa* nest in deciduous woodland, Blean Woods, Kent

Fig. 5.9 *Formica lugubris* nests in the Hope Valley, Derbyshire

In Ireland, *F. lugubris* is found at a few southern sites in Tipperary, Galway and Kerry (Breen, 2014).

Robinson (2011) concluded that *F. lugubris* is also declining, but doing better than its cousin *F. rufa*. She visited 50 historical sites and found 38 still had *F. lugubris* (a 24% decline). Like *F. rufa*, this is a keystone species and it needs to be carefully monitored and protected in the same general way.

5.3 The other ants

This section covers the remaining ants listed in Shirt (1987) and Falk (1991) in their reviews of species under threat.

General environmental issues such as habitat loss and pollution threaten many, if not most, ants. *Formica sanguinea* is under threat from urbanisation, afforestation and agricultural development in its two strongholds of Hampshire in southern England and the Spey Valley in Scotland (Fig. 5.10). It is most commonly found on sandy heaths and at woodland edge in the south of England, and in open woodland in the Spey Valley. Management that encourages the maintenance of these habitats is the best conservation measure for this unique species.

Fig. 5.10 *Formica sanguinea* nest in a rotten tree trunk, near Grantown-on-Spey, Scotland

Loss of heathland is one of the main threats to a variety of species including *Myrmica hirsuta*, *M. karavajevi* and *Strongylognathus testaceus*.

Coastal development, such as the stabilisation of cliffs, threatens another group including *Temnothorax albipennis*, *Myrmica specioides*, *Ponera coarctata*, *P. testacea* and *Solenopsis fugax*.

Agricultural development, repeated fires, recreational pressure, sea-defence construction, invasion by scrub and bracken, afforestation and infilling of old quarries and cuttings are factors that variously threaten many other species including *Myrmica schencki* and *M. specioides*.

The general conclusion is that although special measures such as reintroduction, food supplementation and even captive breeding can be useful, the key to ant conservation is through the careful management of suitable habitats.

6 Identifying ants

6.1 Considerations when identifying ants

This chapter provides keys for identifying all species of ant that are native to the British Isles (Britain, Ireland and the Channel Islands), as well as introduced species that survive in outdoor settings. Heated buildings, including hospitals, glasshouses, and occasionally houses, are sometimes home for exotic (e.g. tropical) species. These are not formally included in the keys but some of the more frequent and widespread exotic species are noted.

Identifying ants can feel daunting. Ants are morphologically rather uniform insects and identification relies on characters seen under high magnification. The addition of over 15 new species to the British Isles list since the first edition of this book has required the inclusion of some subtle characters in the keys. If you are new to ant identification, you may wish to start with the quick key to workers (Key D) and the quick key to nests (Key E). In the main keys (A to C), our aim has been to balance rigorous identification with ease of use. Anatomical terms and the need for precise measurements have been kept to a minimum. Hence, some species will be difficult to distinguish with these keys and may require reference to more specialist works and expertise. For those looking for more precision, recent keys to workers and queens can be found in *The Ants of Central and Northern Europe* (Seifert 2018).

When collecting specimens for identification, a 15× or 20× hand lens is useful for initial identification (for example, with the quick key), but it takes experience to distinguish species this way. For many of the best morphological characters, you will need access to a stereomicroscope – ideally with magnification up to 40× or beyond and good lighting.

Take time to make field observations while collecting, since with experience this will greatly aid subsequent identification. For instance, preserved specimens of *Leptothorax acervorum* and *Myrmica* species may look superficially similar, but in the field they are distinguishable from a distance – a *L. acervorum* worker glides over the ground as though on a cushion of air, whereas a *Myrmica* worker walks in a more measured, 'mechanical' way. Similarly, morphological features for separating the two formicine genera, *Formica* and *Lasius*, can seem subtle, but experience will soon reveal how differently these ants move and behave in life.

Where possible, always collect several workers from a nest, and do not mix collections from different nests, even if the same species is suspected. If you can obtain sexual forms from a nest, these can greatly help to confirm an identification based on workers. For example, workers of *Myrmica sabuleti* and *Myrmica specioides* are quite hard to separate, but their males are very easy to distinguish. Having a nest sample containing different castes will greatly boost the confidence of your identification.

6.2 Anatomical terms

Some formal anatomical terms are used. This is because these terms will be encountered soon enough once the reader delves deeper into more exacting keys. It may help to understand the origins of some unfamiliar terms. An archetypal insect consists of three distinct body sections: head (bearing mouthparts and major sensory organs), thorax (comprising three segments – prothorax, mesothorax, metathorax – each bearing a pair of legs, and the rear two each with a pair of wings) and abdomen (often consisting of a fairly uniform series of segments ending in specialised genitalia and containing major internal organs). In hymenopteran insects, a constriction exists between the first and second abdominal segments such that the first segment appears to be part of the thorax. This segment is called the **propodeum** (Fig. 6.1). Together, the three thoracic segments and the propodeum are termed the **mesosoma**. The propodeum may bear a pair of **propodeal spines** (Fig. 6.2).

A defining feature that distinguishes ants from most bees and wasps is the reduction of the next one or two abdominal segments to form small structures that comprise

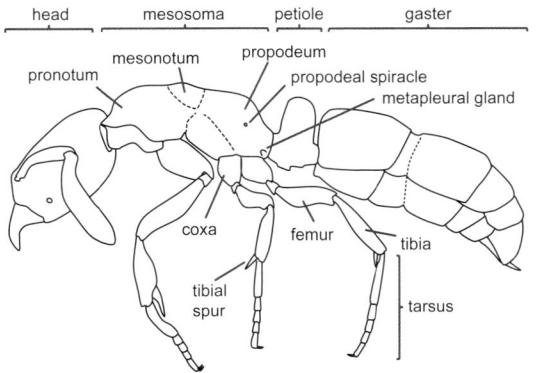

Fig. 6.1 Side view of *Ponera coarctata* worker

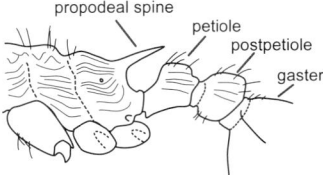

Fig. 6.2 Side view of mid region of *Myrmica ruginodis* worker

the **petiole** and (when present) the **postpetiole**. Collectively, these are the **waist**. The enlarged part of the petiole is called the **node** or **scale**. The remaining segments of the abdomen form the **gaster**.

The mesosoma of the winged castes has a complex structure of articulated plates that will be familiar to those who identify bees and wasps (Figs 6.3, 6.4). The **forewings** are attached at the sides of the **mesonotum** (the top of the mesothoracic segment), which is divided into the **mesoscutum** and **scutellum**. The scutellum is more correctly termed the mesoscutellum, but it is abbreviated here to distinguish it clearly in the keys. The **hindwings** are attached at the sides of the smaller **metanotum**. The **pronotum** at the front of the mesosoma is relatively small.

In contrast to winged ants, the mesosoma of a worker is reduced, with the plates largely fused for strength and rigidity (Fig. 6.1). The mesonotum is strongly reduced and the

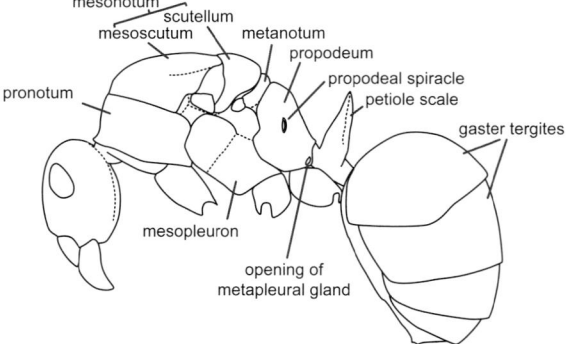

Fig. 6.3 Side view of *Formica pratensis* queen. This queen has shed her wings.

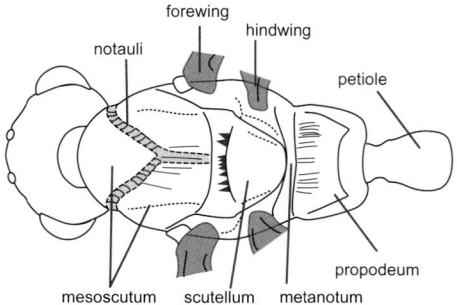

Fig. 6.4 Top view of head and mesosoma of *Myrmica ruginodis* male. The wing bases are shown in grey.

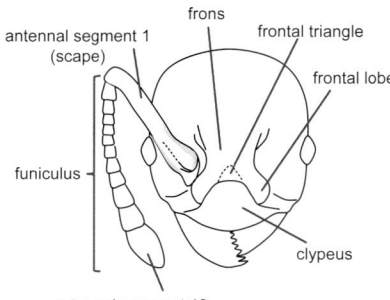

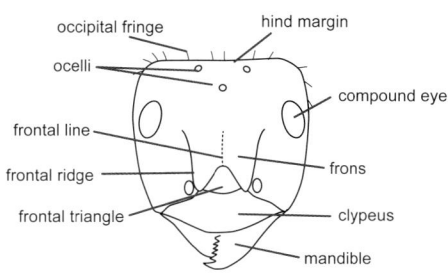

Fig. 6.5 Head of *Myrmica sabuleti* worker

Fig. 6.6 Head of *Formica lugubris* worker

metanotum is rather insignificant. In contrast, the pronotum is larger, accommodating powerful neck muscles.

The head bears many features used in identification, with slightly different terms used for different subfamilies (Figs 6.5, 6.6). The shapes of the **mandibles** and **clypeus** are often important features, as is the **frons** and its associated **frontal triangle**, **frontal line** and **frontal lobes**. The **antenna** consists of an elongated basal segment (the **scape**) with the remaining smaller segments forming the **funiculus**. When the keys refer to number of antennal segments, the scape is included. The underside of the head bears two pairs of mouthpart appendages called **palps**.

6.3 Using the main keys

Before using the main keys (A, B and C), check you have an ant. Usually, it will be sufficient to observe the one- or two-segmented waist that is completely separated from both mesosoma and gaster. Some wasps also appear to have narrow waist segments, but these are either not well separated from the gaster, or are uniformly narrower than shown in Figs 6.1 and 6.2. If you wish to be certain you have an ant, you could look for the opening of the **metapleural gland** on the mesosoma close to the articulation with the petiole (Fig. 6.1).

Then decide which key to use by determining the caste. Workers (Key A) are always wingless. Except in a few species, males are winged, and they are usually distinctive for the obvious genitalia at the end of the gaster (Key C). Queens (Key B, also known as gynes) are initially winged but shed these after mating. They can be distinguished from workers by the wing bases that normally remain, as well as the complex enlarged mesosoma (Fig. 6.3).

antennal segments
When referring to segments by number, the scape is always counted as antennal segment 1, followed by segment 2, and so on.

metapleural gland
A feature unique to ants, which produces antibiotic and antifungal secretions enabling dense colonies in humid nests.

queens and gynes
The term 'queen' is sometimes reserved for reproductive females after mating or colony foundation, while virgin reproductive females are termed 'gynes'.

The keys are arranged in couplets: read both leads of the first couplet, decide which fits better, and then follow the direction to the next numbered couplet and continue until a determination is reached. While gaining experience, weigh up the probability of a correct determination by checking the thumbnail photographs and distribution information given in Chapter 4. For example, upon reaching couplet 5 of Key A you will need to evaluate small differences to decide whether you have a dolichoderine or formicine worker. It is highly likely, however, that you will encounter many *Lasius* and *Formica* workers before you find a *Tapinoma*.

Ants are not particularly colourful, but colour characters can be useful for confirmation and reassurance. Be aware that colours are subjective, and they can vary, especially between living and long-dead specimens. In a few places, body length is used. This is measured from front of head (excluding antenna) to tip of gaster.

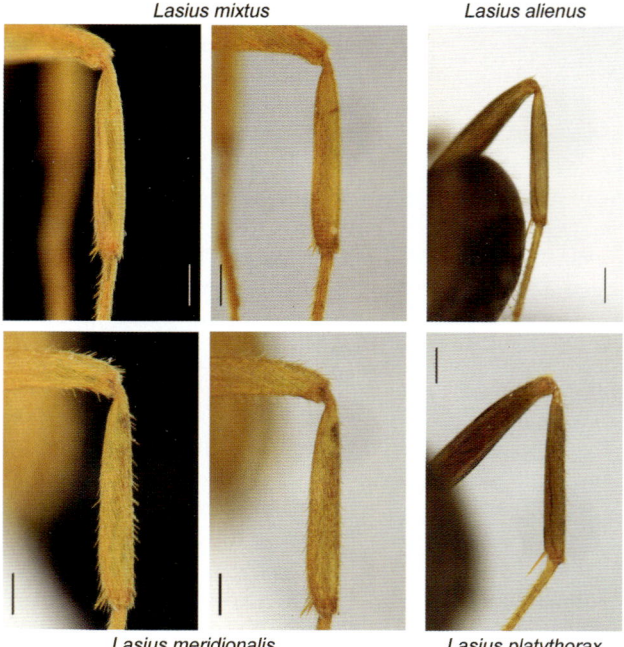

Lasius mixtus　　*Lasius alienus*

Lasius meridionalis　　*Lasius platythorax*

Fig. 6.7 Standing hairs, pubescence, and the importance of illumination. The top-row tibiae of *Lasius* workers have appressed pubescence but no standing hairs except at the tip, compared to the bottom row, which have standing hairs projecting beyond the pubescence. The standing hairs can be surprisingly difficult to see without appropriate illumination or background. Scale bars are 0.2 mm.

pubescence
The silky covering of small hairs appressed to the body surface.

Many couplets depend on assessing the location and number of hairs projecting from the body surface (**erect** or **standing hairs**). These must be distinguished from **pubescence**, which can be profuse in many species (Fig. 6.7). Vary the specimen's angle, illumination, and background exhaustively when looking for hairs, as they can seem all but invisible at some angles (as is demonstrated in Fig. 6.7). Hair characteristics are notoriously variable due to natural variation between workers and to lifetime loss through abrasion. It is essential to consider a sample of workers from a nest (a nest series), for instance when identifying the *Formica* wood ants. Another source of variation is hybridisation between species, which is known or strongly suspected in several *Lasius* species and among the *Formica* wood ants. Sometimes a nest series may not match any species very well.

Key A: Workers

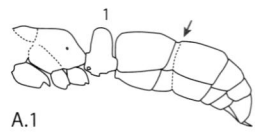

A.1

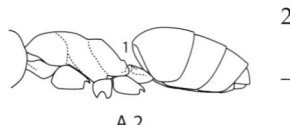

A.2

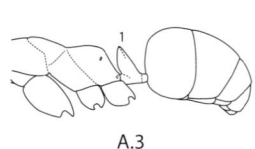

A.3

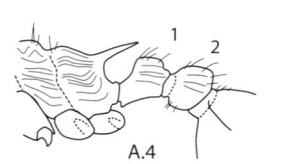

A.4

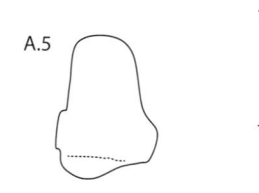

A.5

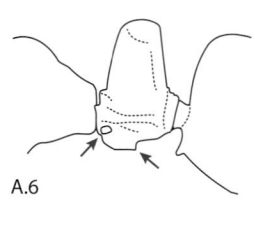

A.6

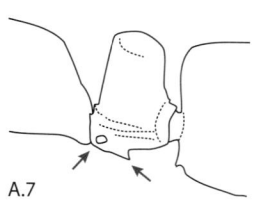

A.7

1　Waist consisting of 1 segment, the petiole; never with spines or teeth on the propodeum (A.1–A.3)　　2

–　Waist consisting of 2 segments, the petiole and postpetiole; usually with a pair of spines or teeth on the propodeum (A.4)　　Myrmicinae 32

2　Gaster with a distinct constriction between the first and second segments (A.1); sting present　Ponerinae 3

–　Gaster with no constriction between the first and second segments (A.2–A.3); sting absent　　5

3　Underside of petiole lacking a backward-pointing tooth or a translucent window (A.5)
Hypoponera punctatissima

This species has occasionally been found outdoors. It and a related species, *Hypoponera ergatandria*, are more often encountered in hothouses etc., and a third, introduced species, *Hypoponera eduardi*, has been seen on a few occasions. The three species are only separable by precise quantitative measurements – if necessary, consult an expert or the keys in Seifert (2018).

–　Underside of petiole with a backward-pointing tooth and a translucent spot or window anteriorly (A.6–A.7)　　genus *Ponera* 4

4　Petiolar node higher and thinner, in dorsal view more elliptical (A.6, A.8); petiolar tooth less pronounced (A.6); ant overall dark brown with dark reddish-brown appendages　*Ponera coarctata*

–　Petiolar node lower and thicker, in dorsal view more circular (A.7, A.9); petiolar tooth more pronounced (A.7); ant overall paler, yellowish-brown; if darker (rarely) then appendages yellowish-brown
Ponera testacea

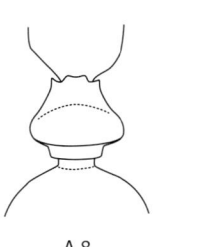

A.8

A.9

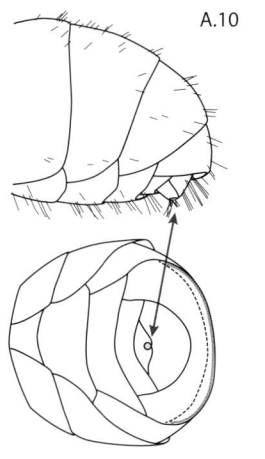

A.10

5   Tip of the gaster with a simple inconspicuous slit across it; upper surfaces of head and mesosoma completely lacking erect hairs   Dolichoderinae 6

–   Tip of gaster with a conical structure terminating in a circular pore surrounded by a collar of hairs (the acidopore) (A.10); upper surfaces of head and mesosoma with at least some, usually with many, erect hairs   Formicinae 8

In a few formicine species, hairs may be confined to the front of the head. Ants of the subfamily Formicinae are much more frequently encountered than those of Dolichoderinae.

6   Petiole surmounted by a scale slanting upwards (similar to A.3), which is visible from above; front margin of clypeus broadly and weakly concave; colour brown   *Linepithema humile*

This is the infamous invasive Argentine ant; it has very occasionally been encountered outdoors in urban areas of Britain. A related species, *Linepithema iniquum*, has been found several times in hothouses.

–   Petiole very low, without a scale, and overhung by the first segment of the gaster so that it is not visible from above (A.2); front margin of clypeus with a distinct notch (A.11–A.12); colour dark brownish-black   genus *Tapinoma* 7

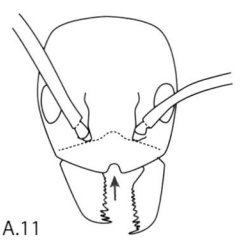

A.11

7   Notch in front margin of clypeus shallow and semicircular (A.11); rear border of head flat or slightly convex   *Tapinoma subboreale*

–   Notch in front margin of clypeus deeper and U-shaped (A.12); rear border of head slightly concave   *Tapinoma erraticum*

Very occasionally *Tapinoma* species from southern Europe have been found outdoors in southern England. These can have workers of noticeably variable size in an individual colony. Several exotic species of *Technomyrmex* are common in hothouses. Workers of this genus closely resemble workers of *Tapinoma* species but can be distinguished by the lack of clypeal notch.

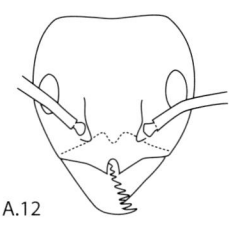

A.12

8   Antenna with 11 segments; a tiny species measuring 1–2 mm (<2.3 mm)   *Plagiolepis pallescens*

Channel Islands only, not recorded so far on mainland Britain or Ireland (although there are occasional sporadic reports of *Plagiolepis* species on the south coast). An exotic species, *Plagiolepis alluaudi*, is frequently found in hothouses of botanical gardens – it is yellow while *P. pallescens* is dark brown.

–   Antenna with 12 segments; larger species generally measuring >2.5 mm   9

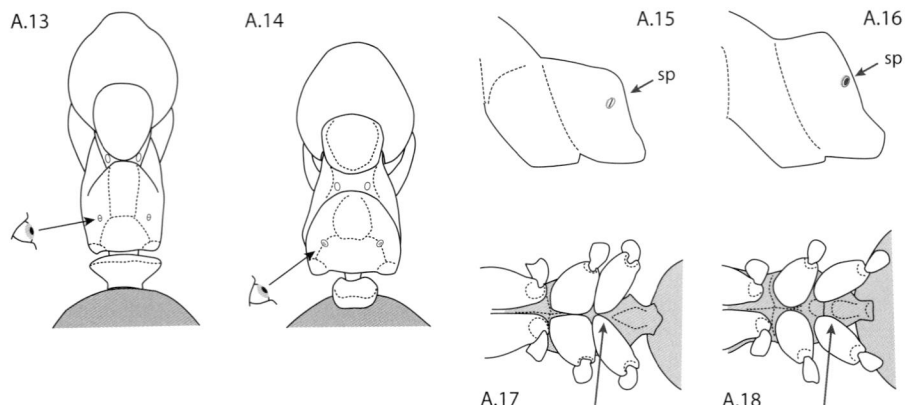

A.13 A.14 A.15 sp A.16 sp

A.17 A.18

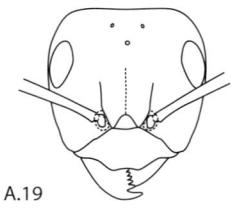

A.19

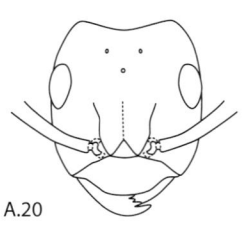

A.20

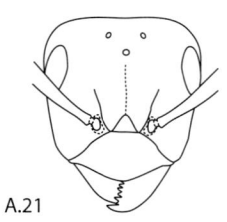

A.21

9   Propodeal spiracle situated on side face of propodeum (A.13); opening of propodeal spiracle elongate and slit-like (although the surrounding rim may appear oval) (A.15 sp); in ventral view, there is no gap between the two hindleg coxae (similar to the mid coxae) (A.17); hind tibia with a double row of bristles on the underside (in addition to those at the tip)

genus *Formica* 10

–   Propodeal spiracle situated on the curved surface between side and rear faces of propodeum (A.14); opening of propodeal spiracle circular or nearly so (A.16 sp); in ventral view, there is a gap between the hind coxae (unlike the mid coxae) (A.18); hind tibia with no rows of bristles on the underside, although there are hairs at the tip and usually abundant pubescence (as in Fig. 6.7)   genus *Lasius* 20

The spiracle shape character requires some care: be sure to observe the spiracle from a viewpoint perpendicular to the surface on which it is located as shown in A.13 and A.14, and assess the shape of the opening rather than the surrounding rim.

10   Front margin of clypeus with a notch in the middle (A.19)   *Formica sanguinea*

–   Front margin of clypeus without a notch (A.20–A.21)   11

11   Head strongly concave behind (A.20)   *Formica exsecta*

–   Head not strongly concave behind (A.21)   12

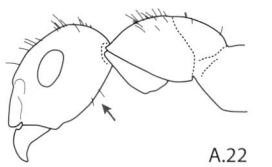

A.22

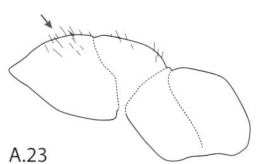

A.23

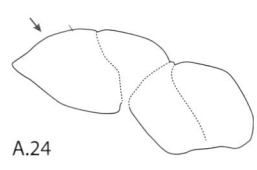

A.24

12   Body (excluding legs) unicoloured dark brown or black, never with any red colour   13

–   Body dark brown with lighter red or yellowish-red areas, and therefore often appearing bicoloured. Red always present but variable in extent – it can be extensive on mesosoma, petiole scale and head, or sometimes confined to patches on the sides of pronotum and mesonotum   15

13   Shining due to sparse pubescence on head and gaster; lower surface of head with two long hairs (these are occasionally missing in abraded specimens) (A.22)
      Formica picea

–   Not strongly shining, the pubescence being more dense and silky; lower surface of head without hairs   14

14   Top of pronotum with numerous short, erect hairs; femur of middle leg with a few long hairs beneath (A.23)   *Formica lemani*

–   Top of pronotum with few, if any, erect hairs (0–3); femur of middle leg without long hairs (A.24)
      Formica fusca

15   Frontal triangle finely roughened and not reflecting light (similar to frons)   16

–   Frontal triangle unsculptured or with very fine pits, and shining (in clear contrast to frons)   17

16   Pronotum and mesonotum both with numerous erect hairs (similar to A.23)   *Formica rufibarbis*

      Rare and highly endangered on mainland Britain. Also well known from the Isles of Scilly.

–   Pronotum and mesonotum usually without erect hairs (but occasionally up to three on the former) (similar to A.24)   *Formica cunicularia*

17   Hind margin of head with a fringe of few-to-many standing hairs (occipital fringe, see Fig. 6.6), sometimes reaching round at least to the level of the compound eyes; compound eye with minute erect hairs between facets (use high magnification and vary the lighting)   18

–   Hind margin of head without an occipital fringe; compound eye usually with no erect hairs
      Formica rufa

      Research suggests that the species in Britain long referred to as *F. rufa* is in fact a stable hybrid of two non-British species, *Formica polyctena* and *F. rufa* (Seifert *et al.* 2010).

A.25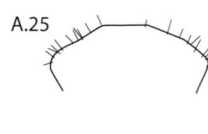

A.26

A.27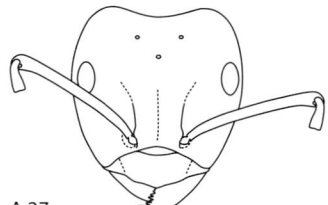

A.28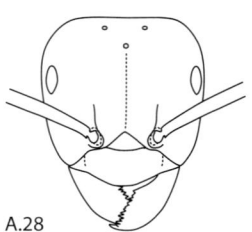

18   Head with frons matt due to extremely fine micro-sculpture; frontal triangle more strongly pitted

Formica pratensis

Channel Islands only, extinct on mainland Britain.

–   Frons weakly shining and with fine pits scattered over it; frontal triangle with only a few fine pits   19

19   Generally hairier: upper margin of petiole scale above the widest part with >8 projecting hairs (A.25); in profile, mesosoma usually with numerous standing hairs (typically >15 projecting above the mesonotum); occipital fringe well developed, extending down the sides of the head to the compound eyes

Formica lugubris

–   Generally less hairy: upper margin of petiole scale above the widest part with <8 projecting hairs (A.26); in profile, mesosoma usually with sparse standing hairs (typically <10 projecting above the mesonotum); occipital fringe sparse, rarely extending as far as the compound eyes   *Formica aquilonia*

Individuals of these species show wide variation in hairiness, but if several workers are examined for an average view, the identification is usually clear.

20   Head strongly concave behind, so as to be heart-shaped (A.27); colour glossy black   *Lasius fuliginosus*

–   Head rounded or at most weakly concave behind (A.28); colour dark brown, light brown or yellow, and diffusely shining at best, not glossy   21

21   Colour predominantly dark, with at least the gaster dark brown; maxillary palps (the longer of the two pairs of appendages on the mouthparts) relatively long and conspicuous: when laid back under the head they reach the level of the upper margin of the eye (A.29)   22

–   Colour yellow or brownish-yellow; maxillary palps relatively short: when laid back under the head they distinctly fail to reach the level of the upper margin of the eye (A.30)   28

A.29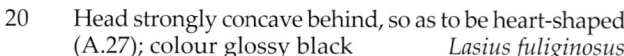

A.30

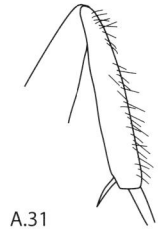

A.31

22   Antennal scape and outer edge of middle and hind tibia with numerous erect or slightly slanting hairs (A.31)   23

–   Antennal scape and outer edge of middle and hind tibia each with at most 1–2 such hairs, and often without any hairs at all   25

23   Mesosoma reddish, as is the clypeus and underside of the head, contrasting with dark gaster

Lasius emarginatus

Native to Channel Islands and recently established in southern England.

–   Body entirely dark brown   24

24   With the head in profile, the clypeus more convex and slightly protruding (shaped like an aquiline nose) (A.32); short hairs on clypeus dense (A.33)

Lasius niger

–   With the head in profile, the clypeus flatter (A.34); short hairs on clypeus sparse such that the underlying shining surface is clearly visible (A.35)

Lasius platythorax

Adjust the lighting to see the clypeal hairs.

25   Bicoloured, the gaster much darker than the light brown mesosoma; petiole scale with a broad, shallow V-shaped notch at the top (A.36); rear corners of head without projecting hairs   *Lasius brunneus*

–   Unicoloured, the gaster and mesosoma the same colour or nearly so; petiole scale usually rounded with at most a slight narrow indentation at the top (A.37); rear corners of head with projecting hairs   26

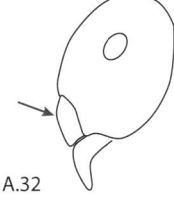

A.32

A.33

A.34

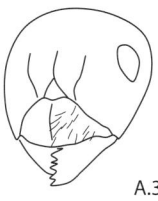

A.35

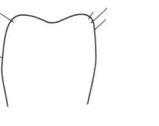

A.36   A.37

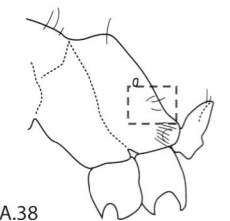

A.38

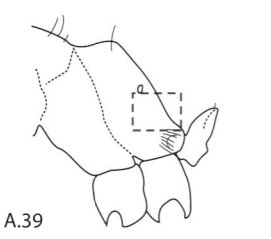

A.39

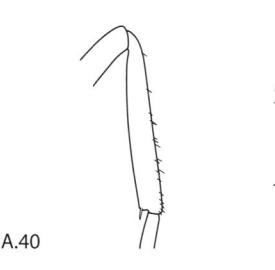

A.40

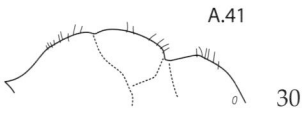

A.41

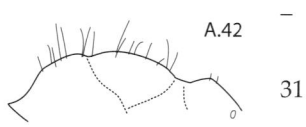

A.42

26   Mandibles commonly with 7 teeth (examine several workers); mandibles paler than rest of head, the same colour as the appendages   *Lasius neglectus*

Invasive species in urban parks, botanical gardens, etc.

–   Mandibles commonly with 8 teeth; mandibles nearly as dark as rest of head   27

27   Area between the propodeal spiracle and the metapleural gland opening with 3–5 short, erect hairs (A.38, boxed region)   *Lasius psammophilus*

–   Area between the propodeal spiracle and the metapleural gland opening with 0 short erect hairs (rarely 1) (A.39, boxed region)   *Lasius alienus*

28   Antennal scape and hind tibia with pubescence, but without erect or slightly slanting hairs (as defined in Fig. 6.7) except at tip; sides of head between eye and mandible lacking erect hairs   29

–   Antennal scape and hind tibia with both pubescence and short erect or slightly slanting hairs (A.40); sides of head between eye and mandible with erect hairs   30

29   Erect hairs present on underside of head; hairs projecting from mesosoma all shorter than the length of the eye (A.41)   *Lasius mixtus*

–   No erect hairs on underside of head; many hairs projecting from mesosoma as long as or longer than length of the eye (A.42)   *Lasius flavus*

A closely related species, *Lasius myops*, has been reported (rarely) in the Channel Islands and possibly mainland Britain. Distinguishing it from *L. flavus* requires careful measurement of eye size (Seifert 2018).

30   Fore tibia without erect or slightly slanting hairs; those of hind tibia sparse   *Lasius sabularum*

–   Fore tibia with erect or slightly slanting hairs; those of hind tibia numerous   31

31   Scape and tibia somewhat flattened in cross-section   *Lasius meridionalis*

–   Scape and tibia almost circular in cross-section   *Lasius umbratus*

Workers of *L. sabularum*, *L. meridionalis* and *L. umbratus* are very difficult to distinguish without extensive precise measurements; if necessary, consult an expert.

A.43

A.44

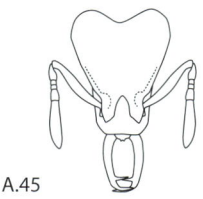

A.45

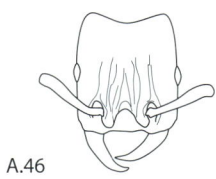

A.46

32   Spines of propodeum narrow, with sharply pointed tips, and often long (A.43)   genus *Myrmica* 47

–   Spines of propodeum different from above – either short and triangular, longer but blunt-tipped, or absent (A.44)   33

This couplet is not definitive, but usually enables *Myrmica*, the commonest myrmicine ants, to be keyed out early for convenience. If in doubt, take the second branch (which eventually leads again to *Myrmica*). See Fig. 6.8 for representative photographic images of spines.

33   Antenna with 6 segments; head oddly shaped with projecting mandibles resembling tongs (A.45)

Strumigenys perplexa

A tiny introduced species discovered in the Channel Islands in 2020 (Hamer *et al.* 2021).

–   Antenna with at least 10 segments; head shape and mandibles different from above   34

34   Mandibles long, slender and evenly curved like a scythe blade (A.46)   *Strongylognathus testaceus*

Found only in nests of its host, *Tetramorium caespitum*.

–   Mandibles with a toothed cutting edge (as in A.52, A.77), with at least four distinct teeth   35

Fig. 6.8 Propodeal spines. Most species of *Myrmica*, like *M. ruginodis*, are readily recognised by their long, narrow and vicious-looking spines. In other myrmicine genera, such as *Leptothorax* or *Tetramorium*, the spines are typically less robust and often triangular. The common species, *Myrmica rubra*, can be confusing as it has shorter spines than most *Myrmica*, but they are still relatively narrow and pointed.

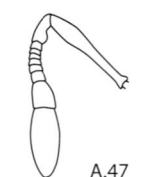

A.47

A.48

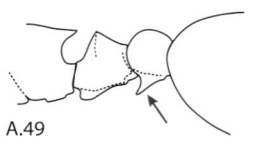

A.49

35   Antenna with 10 segments, the two at the tip forming a club (A.47); propodeum unarmed, without a pair of teeth or spines (A.48)   *Solenopsis fugax*

–   Antenna with 11 or 12 segments; propodeum armed with a pair of teeth or spines (e.g. A.49, A.54, A.62)   36

Several introduced *Monomorium* species can be found in heated buildings, including the notorious pharaoh ant, *M. pharaonis*; these ants would key to this couplet, but have 12 antennal segments and an unarmed propodeum.

36   Postpetiole with a projection below (A.49); upper surfaces of head and mesosoma glossy and mirror-like   *Formicoxenus nitidulus*

Living in the mounds of *Formica* wood ants only. Both the workers and the remarkably worker-like males will key out here (antenna with 11 and 12 segments respectively).

–   Postpetiole without projection below (as in A.50, A.68); upper surfaces of head and mesosoma at least partly sculptured and often strongly wrinkled   37

37   Antenna with 11 segments   *Leptothorax acervorum*

–   Antenna with 12 segments   38

38   Lower surface of head with a sharp ridge along each side; petiole almost square or pentagonal in profile (A.50); front margin of clypeus with two prominent teeth, sometimes with an additional small one in the middle   *Myrmecina graminicola*

–   Lower surface of head without a pair of ridges; petiole not square, at least narrowing in front (e.g. A.61, A.68); front margin of clypeus not toothed   39

39   Front of mesosoma with angular corners (square-shouldered) (A.51); sides of the clypeus raised into sharp semicircular ridges bordering the deep antennal sockets (A.52)   *Tetramorium caespitum*

A closely related species, *Tetramorium impurum*, has been reported from the Channel Islands. In contrast to *T. caespitum*, workers lack shiny areas on petiole and postpetiole, but they are difficult to separate without precise measurements. Several exotic species of *Tetramorium* are common in hothouses. In these species, the head has frontal ridges extending up to its rear corners.

A.50

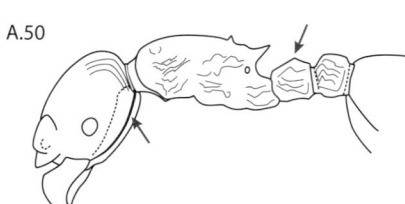

A.51

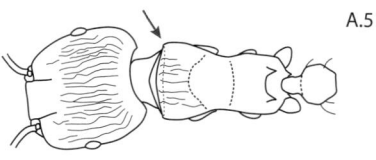

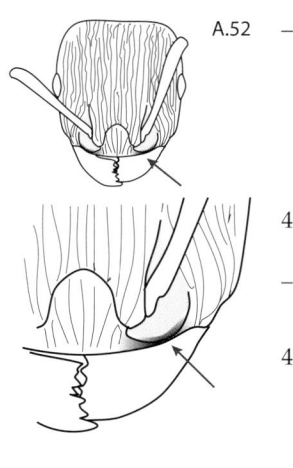

A.52   –   Front of mesosoma smoothly curved into sides; sides of the clypeus may be narrow but are not raised into sharp ridges, instead smoothly transitioning into front of clypeus (as in A.77–A.78)   40

The clypeal character requires some experience. If in doubt, go with the mesosoma character.

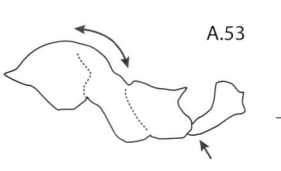

40   Petiole elongate in front and without a small tooth underneath (A.53–A.54)   41

–   Petiole narrowed in front but not elongate, and with a small tooth underneath (e.g. A.61, A.68)   43

41   Middle portion of clypeus without a pair of ridges; mesonotum domed and strongly curved into the propodeum (A.53)   *Aphaenogaster subterranea*

Channel Islands only, not found on mainland Britain or Ireland. In 2024, the presence of *Messor capitatus* was confirmed on Guernsey. It would key to this couplet, but is black rather than brown and has strongly polymorphic workers (pp. 33, 82).

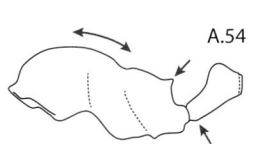

A.53

–   Middle portion of clypeus with a pair of prominent ridges (A.55–A.56); mesonotum not strongly domed and gently curved into the propodeum (A.54)   genus *Stenamma* 42

A.54

42   Width of central smooth area of frons (frontal triangle) about one third of the total frons width (A.55); viewed from above, front of petiole has a prominent nub on each side (A.57)   *Stenamma debile*

–   Width of central smooth area of frons (frontal triangle) less than one third of the total frons width (A.56); viewed from above, front of petiole lacks side nubs (A.58)   *Stenamma westwoodii*

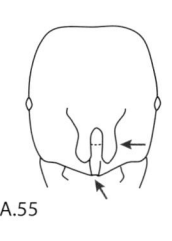

A.55

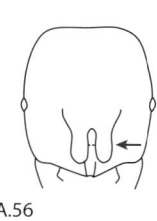

A.56

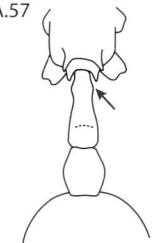

A.57

A.58

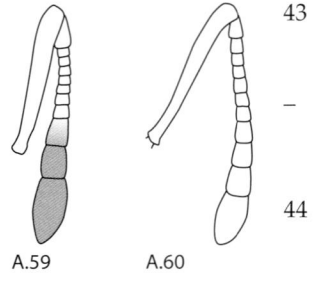

A.59   A.60

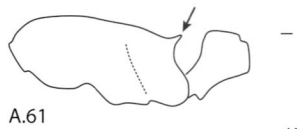

A.61

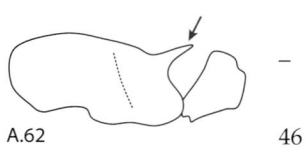

A.62

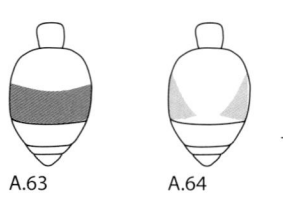

A.63   A.64

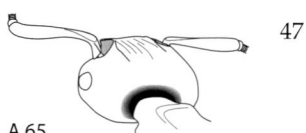

A.65

43   The last three antennal segments nearly as long as the remainder of the funiculus (A.59); mandibles with 4–6 teeth   genus *Temnothorax* 44

–   The last three antennal segments distinctly shorter than the remainder of the funiculus (A.60); mandibles with 7–10 teeth   genus *Myrmica* 47

44   Antenna with terminal three segments the same pale brown colour as the remaining segments; gaster with a very broad dark band on the first gastral segment and narrower bands on the second and third segments
     Temnothorax nylanderi

–   Antenna with terminal three segments distinctly darker than the remaining segments (A.59); banding of gastral segments not as above   45

45   Spines on propodeum short and straight (A.61); much of head darker than the mesosoma
     Temnothorax albipennis

–   Spines on propodeum longer and curved (A.62); head largely as pale as the mesosoma except at the front   46

46   Top of the head in middle with sculpture consisting of an irregular pattern with wrinkles; orange-yellow with a neat dark band across first segment of gaster (A.63)   *Temnothorax unifasciatus*

Channel Islands only with a single record on mainland Britain, presumably as an introduction.

–   Top of the head in middle with very fine longitudinal lines discernible; orange-yellow with vague darker patches on front of head and sides of first gastral segment (A.64)   *Temnothorax interruptus*

47   Viewed from the rear (as shown in A.65), basal part of antennal scape curved, sometimes strongly so; the curved part lacks ridges, flattened areas or outgrowths, and therefore remains approximately circular in cross-section throughout the curve (A.66–A.67)   48

–   Viewed from the rear (as shown in A.65), basal part of antennal scape distinctly angled on the outside of an abrupt bend; the bent part with ridges, flattened areas or outgrowths (A.72–A.76)   50

Distinguishing *Myrmica* species relies heavily on the shape of the base of the antennal scape (i.e. towards its attachment to the head). The scape should be manipulated to extend laterally from the head. Rear view refers to looking from the rear of the head such that the funiculus points directly away from the observer, while front view refers to looking at the head face on. Photographs of front and rear views of the scape are shown in Fig. 6.9.

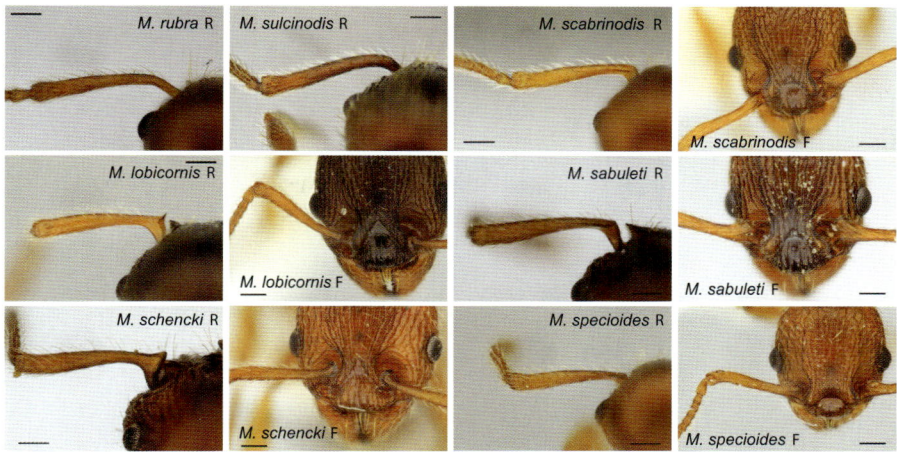

Fig. 6.9 Antennal scapes of *Myrmica* species. The shape of the scape base is an important character and should be viewed from both the rear (R) and the front (F). Scale bars are 0.2 mm.

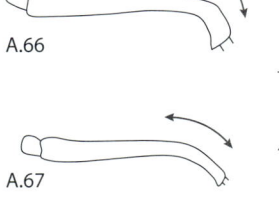

48 Scape stout and strongly curved near the base, nearly forming a right angle (A.66); frontal triangle extensively sculptured   *Myrmica sulcinodis*

– Scape slender and gently curved in an obtuse angle (A.67); frontal triangle largely smooth, shining   49

49 Upper surface of petiole curved into the articulation with postpetiole (A.68); sides of petiole and postpetiole weakly wrinkled; spines on propodeum shorter than the distance between their tips (A.68, A.70)
    Myrmica rubra

– Upper surface of petiole abruptly angled before meeting postpetiole (A.69); sides of petiole and postpetiole strongly wrinkled; spines on propodeum long, about as long as the distance between their tips (A.69, A.71)   *Myrmica ruginodis*

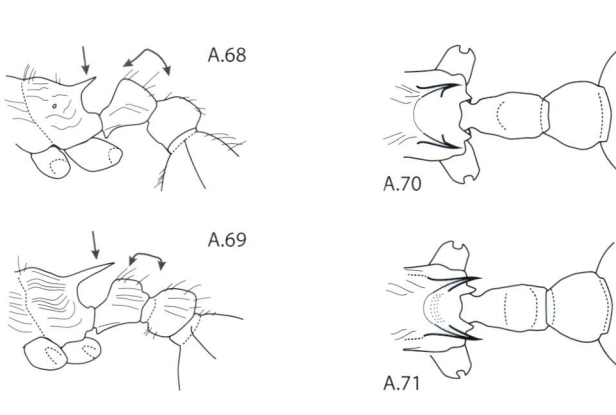

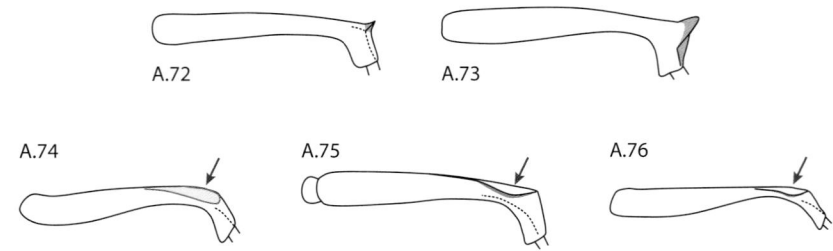

50 Bend of scape with an outgrowth projecting transversely across its outer corner, appearing tooth-like or flange-like when viewed from the rear (A.72–A.73) 51

– Bend of scape without a transverse flange or tooth, but with a ridge or outgrowth running along the long axis of the scape (A.74–A.76) 52

51 Outgrowth of scape bend is narrow and tooth-like (A.72, A.77); frontal lobes wide in comparison to the frontal triangle (A.77); body colour dark reddish-brown, sometimes with lighter mesosoma
Myrmica lobicornis

– Outgrowth of scape broad and flange-like, wrapping round bend (A.73, A.78); frontal lobes narrow in comparison to the frontal triangle (A.78); body colour clear red *Myrmica schencki*

52 Postpetiole wider than high; head sometimes with traces of ocelli (see Fig. 6.6) *Myrmica hirsuta*

A rare species found only in the nests of its host *Myrmica sabuleti*. This is not a true worker caste but a wingless, ergatoid female and such individuals are very infrequent in parasitised nests.

– Postpetiole higher than wide; head never with traces of ocelli 53

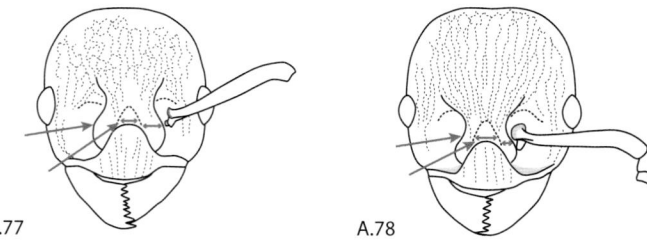

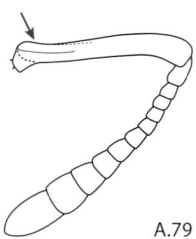

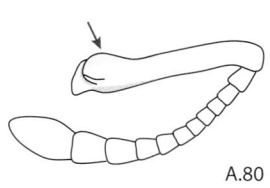

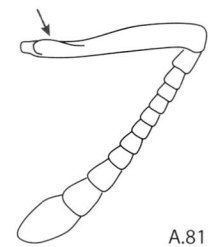

A.79

A.80

A.81

53   Clypeus with front margin slightly concave in middle   *Myrmica vandeli*

Rare and little known in Britain.

–   Clypeus with front margin straight or convex   54

54   Scape with a low ridge extending longitudinally from the bend. The ridge is inconspicuous but borders a flattened area of the surface that is angled such that it is visible not only from the front (A.79) but also when viewed strictly from the rear (A.65, A.74)   *Myrmica scabrinodis*

–   Scape with a sharp-edged projecting ridge extending longitudinally from the bend. The flattened surface of the projection faces the observer when viewed from the front (A.80–A.81), but is barely visible when viewed strictly from the rear (as its sharp edge projects towards the observer) (A.75–A.76)   55

55   Petiole with an abrupt angle between upper margin and rear margin (A.82); outgrowth of scape usually prominent (A.80)   *Myrmica sabuleti*

A montane form of *M. sabuleti* with an even larger, upswept scape outgrowth has been considered a separate species, *Myrmica lonae*, although DNA analysis has cast doubt on this. This form has been found in Scotland (see photograph on p. 154).

–   Petiole with upper margin evenly curved to its join with the postpetiole (A.83); outgrowth of scape sharp but less prominent (A.81)   *Myrmica specioides*

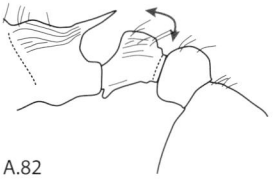

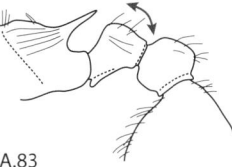

A.82

A.83

Key B: Queens

The characters used to identify queens are often similar to those for workers of the same species, and so reference is made to some figures in the worker key.

1   Waist consisting of 1 segment, the petiole (B.1–B.3)   2

–   Waist consisting of 2 segments, the petiole and postpetiole (B.4)   Myrmicinae 32

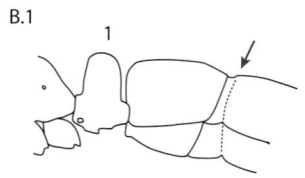

B.1

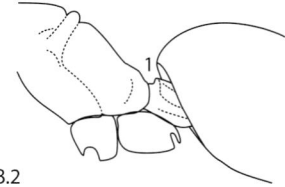

B.2

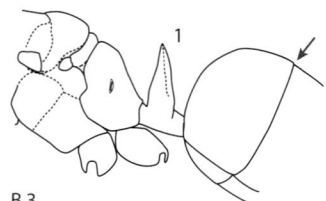

B.3

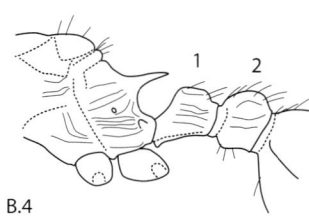

B.4

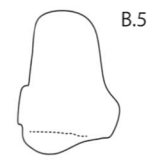

B.5

2   Gaster with a distinct constriction between the first and second segments (B.1); sting present   Ponerinae 3

–   Gaster with no constriction between the first and second segments (B.3); sting absent   5

3   Underside of petiole lacking a backward-pointing tooth or a translucent window (B.5)

Hypoponera punctatissima

This species has occasionally been found outdoors. It and a related species, *Hypoponera ergatandria*, are more often encountered in hothouses etc.; they are only separable by precise quantitative measurements – if necessary, consult an expert.

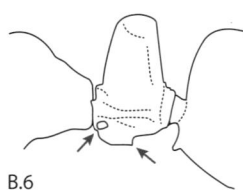

B.6

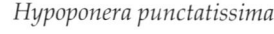

–   Underside of petiole with a backward-pointing tooth and a translucent spot or window anteriorly (B.6–B.7)   genus *Ponera* 4

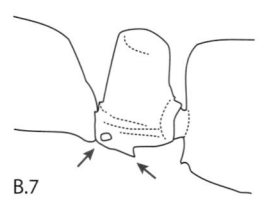

B.7

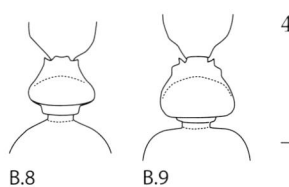

B.8   B.9

4   Petiolar node higher and thinner (B.6, B.8); petiolar tooth less pronounced (B.6); ant overall dark brown with dark reddish-brown appendages
Ponera coarctata

–   Petiolar node lower and thicker (B.7, B.9); petiolar tooth more pronounced (B.7); ant overall paler, yellowish-brown; if darker (rarely) then appendages yellowish-brown   *Ponera testacea*

5   Tip of the gaster with a simple inconspicuous slit across it; upper surfaces of head and mesosoma completely lacking erect hairs   Dolichoderinae 6

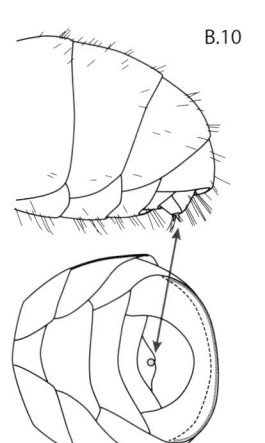

B.10

–   Tip of gaster beneath with a conical structure terminating in a circular pore surrounded by a collar of hairs (the acidopore) (B.10); upper surfaces of head and mesosoma with at least some, usually many, erect hairs   Formicinae 8

In a few formicine species, hairs may be confined to the front of the head. Ants of the subfamily Formicinae are much more frequently encountered than those of the Dolichoderinae.

6   Petiole surmounted by a scale slanting upwards (similar to B.3), which is visible from above; front margin of clypeus broadly and weakly concave; colour brown   *Linepithema humile*

This is the infamous invasive Argentine ant; it has very occasionally been found outdoors in urban areas of Britain. A related species, *Linepithema iniquum*, has been found several times in hothouses.

–   Petiole low, without a scale, and overhung by the first segment of the gaster (B.2) so that it is not visible from above; front margin of clypeus with a distinct notch (B.11–B.12); colour dark brownish-black
genus *Tapinoma* 7

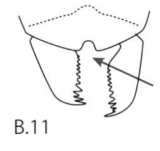

B.11

7   Notch in front margin of clypeus shallow and semicircular (B.11); queen smaller (4.5–5.5 mm)
Tapinoma subboreale

–   Notch in front margin of clypeus deeper and U-shaped (B.12); queen usually larger   *Tapinoma erraticum*

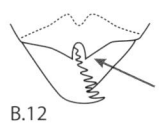

B.12

8   Antenna with 11 segments; a tiny species measuring <4 mm   *Plagiolepis pallescens*

Channel Islands only, not present on mainland Britain or Ireland (although there are occasional sporadic reports of *Plagiolepis* species on the south coast). An exotic species, *Plagiolepis alluaudi,* is frequently found in hothouses of botanical gardens.

–   Antenna with 12 segments; larger species generally measuring >5 mm   9

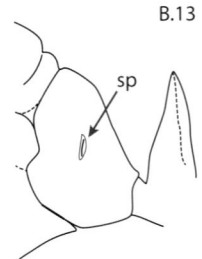
B.13

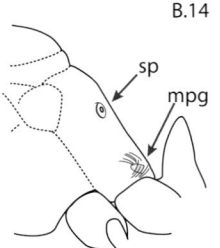
B.14

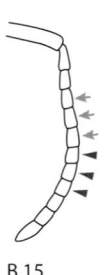
B.15

B.16

9 Opening of propodeal spiracle elongate, slit-like (although the surrounding rim appears oval) (B.13 sp); antennal segments 4, 5 and 6 each slightly but perceptibly elongated compared to segments 7, 8 and 9 (B.15); hind tibia with a double row of bristles present on the underside (in addition to those at the tip) genus *Formica* 10

– Opening of propodeal spiracle circular or nearly so (B.14 sp); antennal segments 4, 5 and 6 similar lengths to segments 7, 8 and 9 (B.16); hind tibia with no rows of bristles on the underside, although there are hairs at the tip and usually abundant pubescence (as in Fig. 6.7) genus *Lasius* 20

10 Front margin of clypeus with a notch in the middle (B.17) *Formica sanguinea*

– Front margin of clypeus without a notch in the middle (B.18–B.19) 11

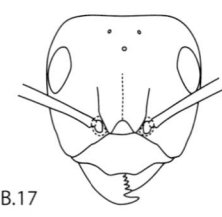
B.17

11 Head strongly concave behind (B.18) *Formica exsecta*

– Head not strongly concave behind (B.19) 12

12 Body (excluding legs) uniformly brownish-black or black 13

– Body bicoloured brownish-black and reddish; red can be extensive or confined to parts of the mesosoma 15

13 Shining due to sparse pubescence; lower surface of head with 2–3 long hairs (these are occasionally missing in abraded specimens) *Formica picea*

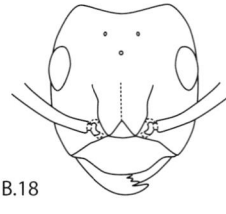
B.18

– Head and mesosoma with silky pubescence and not strongly shining (although gaster may be shining); lower surface of head without hairs 14

14 Row of erect hairs on top of pronotum extending round to wing bases; femur of middle leg with a few long hairs beneath *Formica lemani*

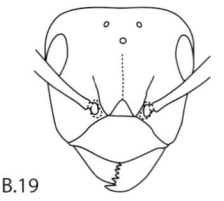
B.19

– Row of erect hairs on top of pronotum confined to the front portion; femur of middle leg without long hairs *Formica fusca*

15 Frontal triangle finely rough, dull, not reflecting light (similar to frons) 16

–   Frontal triangle sometimes with very fine punctures but clearly smoothly shining and mirror-like (in contrast to frons) 17

16 Top of propodeum with a few long hairs; mesoscutum usually bicoloured *Formica rufibarbis*

Rare and highly endangered on mainland Britain. Also well known from the Isles of Scilly.

–   Top of propodeum without long hairs; mesoscutum usually uniformly dark, occasionally reddish around wing bases *Formica cunicularia*

17 Gaster, scutellum and mesoscutum each matt and dull *Formica pratensis*

Channel Islands only, extinct on mainland Britain.

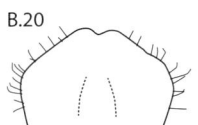
B.20

–   Gaster and scutellum shining, in contrast to more densely sculptured mesoscutum 18

18 Upper margin of petiole (B.20) and propodeum sides with long hairs *Formica lugubris*

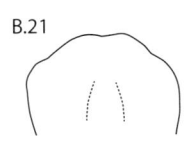
B.21

–   Upper margin of petiole (B.21) and propodeum sides bare or with short, sparse hairs 19

19 Compound eye with short erect hairs between the facets (use high magnification and vary the lighting) *Formica aquilonia*

–   Compound eye bare or with a few microscopic hairs *Formica rufa*

Research suggests that the species known in Britain as *F. rufa* is in fact a stable hybrid of two non-British species, *Formica polyctena* and *F. rufa* (Seifert *et al.* 2010).

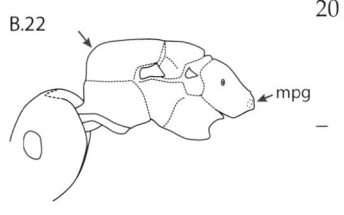
B.22

20 Colour glossy black; opening of metapleural gland lacks collar of guard hairs (B.22 mpg); mesoscutum high, with front surface appearing perpendicular to top surface (B.22) *Lasius fuliginosus*

–   Colour dark brown to yellowish-brown, at most diffusely shining; opening of metapleural gland surrounded by guard hair collar (e.g. B.25 mpg); mesoscutum low and evenly curved into pronotum (e.g. B.25–B.26) 21

21   Maxillary palps (the longer of the two pairs of appendages on the mouthparts) relatively long and conspicuous: when laid back under the head they reach the level of the eye (similar to A.29)   22

–   Maxillary palps relatively short: when laid back under the head they fail to reach the level of the eye (similar to A.30)   28

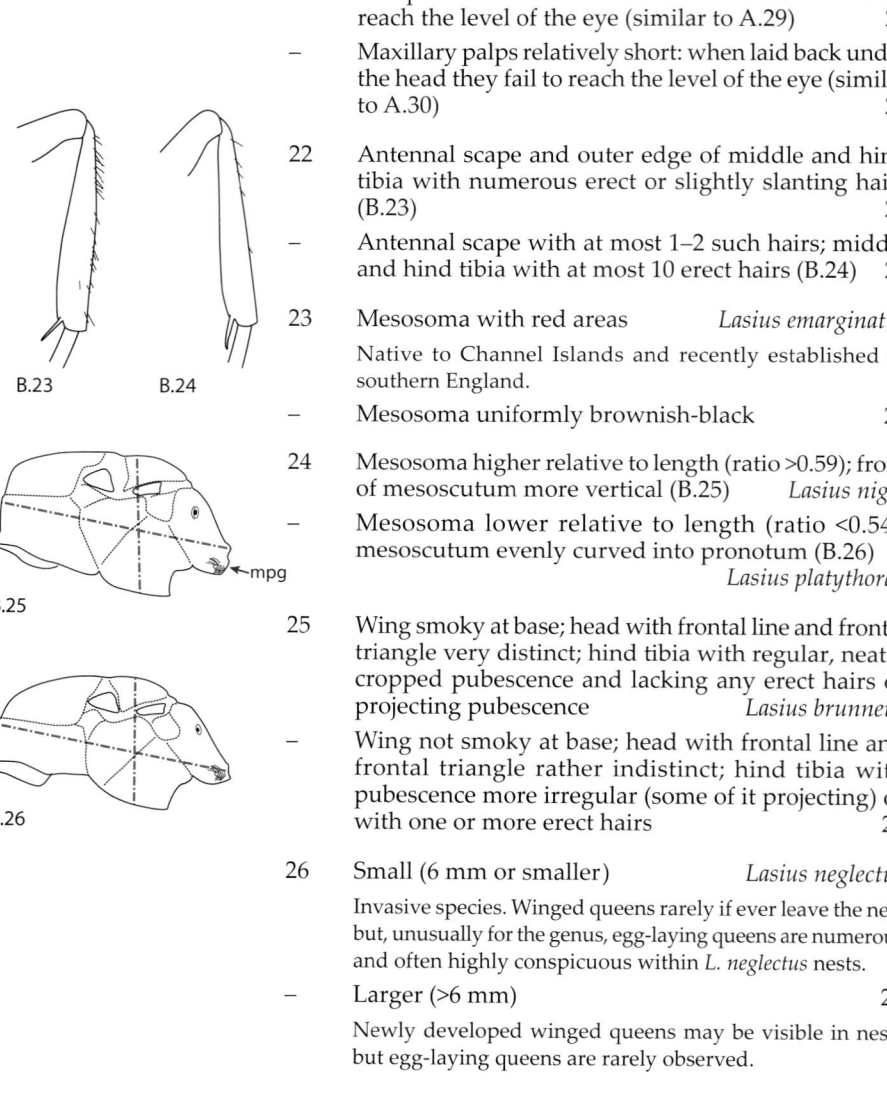

B.23   B.24

B.25

B.26

22   Antennal scape and outer edge of middle and hind tibia with numerous erect or slightly slanting hairs (B.23)   23

–   Antennal scape with at most 1–2 such hairs; middle and hind tibia with at most 10 erect hairs (B.24)   25

23   Mesosoma with red areas   *Lasius emarginatus*

Native to Channel Islands and recently established in southern England.

–   Mesosoma uniformly brownish-black   24

24   Mesosoma higher relative to length (ratio >0.59); front of mesoscutum more vertical (B.25)   *Lasius niger*

–   Mesosoma lower relative to length (ratio <0.54); mesoscutum evenly curved into pronotum (B.26)   *Lasius platythorax*

25   Wing smoky at base; head with frontal line and frontal triangle very distinct; hind tibia with regular, neatly cropped pubescence and lacking any erect hairs or projecting pubescence   *Lasius brunneus*

–   Wing not smoky at base; head with frontal line and frontal triangle rather indistinct; hind tibia with pubescence more irregular (some of it projecting) or with one or more erect hairs   26

26   Small (6 mm or smaller)   *Lasius neglectus*

Invasive species. Winged queens rarely if ever leave the nest but, unusually for the genus, egg-laying queens are numerous and often highly conspicuous within *L. neglectus* nests.

–   Larger (>6 mm)   27

Newly developed winged queens may be visible in nests but egg-laying queens are rarely observed.

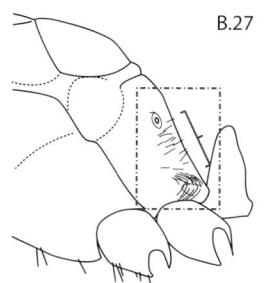

B.27

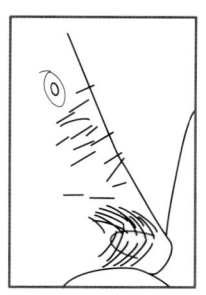

B.28

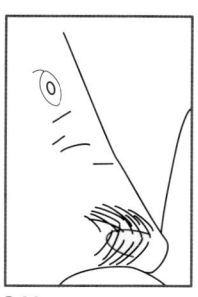

B.29

27 Area between the propodeal spiracle and the
    metapleural gland opening (B.27) with 6–20 short, erect
    hairs (B.28); outer edge of hind tibia with projecting
    hairs (up to 10) (like B.32)   *Lasius psammophilus*

–   Area between the propodeal spiracle and the
    metapleural gland opening with up to 6 short, erect
    hairs (B.29); outer edge of hind tibia usually without
    projecting hairs   *Lasius alienus*

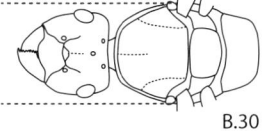

B.30

28 Head narrower than mesosoma (B.30); body length
    larger, about 3–4 times the length of the workers of
    the same species   *Lasius flavus*

    A closely related species, *Lasius myops*, has been reported
    (rarely) in the Channel Islands and possibly mainland Britain.
    The head of *L. flavus* queen has orange-brown patches only on
    the side/underside next to the mandibles, whereas *L. myops*
    has the entire underside of the head orange-brown.

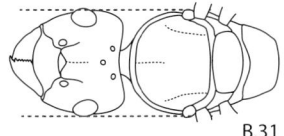

B.31

–   Head about as wide as mesosoma (B.31); body length
    distinctly smaller, about twice the length of the workers
    of the same species   29

29 Hind tibia with profuse pubescence but without
    standing hairs   *Lasius mixtus*

–   Hind tibia with pubescence and at least a few (often
    many) short, standing hairs (B.32–B.34)   30

    Be sure to distinguish standing hairs from the abundant,
    short pubescence found in most species (see Fig. 6.7).

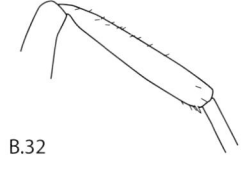

B.32

30 Hind tibia with few standing hairs (B.32) and front
    tibia with few or no standing hairs *Lasius sabularum*

–   Hind tibia (B.33–B.34) and front tibia both with
    numerous standing hairs   31

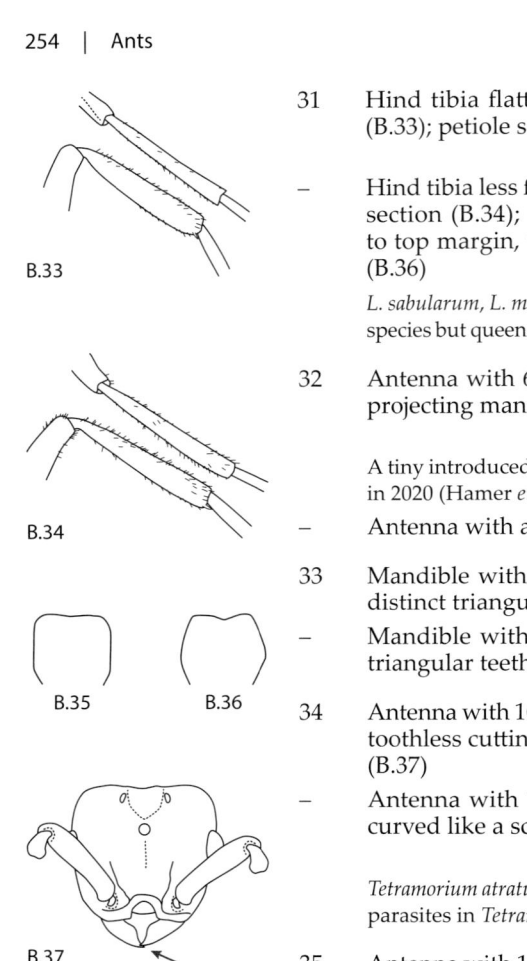

B.33

B.34

B.35   B.36

B.37

B.38

31   Hind tibia flattened, about twice as wide as deep (B.33); petiole scale almost rectangular (B.35)
Lasius meridionalis

–   Hind tibia less flattened and more roundish in cross-section (B.34); petiole scale with sides converging to top margin, which is usually concave or notched (B.36)   *Lasius umbratus*

L. sabularum, L. meridionalis and *L. umbratus* are very similar species but queens provide the best characters for separation.

32   Antenna with 6 segments; head oddly shaped with projecting mandibles resembling tongs (as A.45)
Strumigenys perplexa

A tiny introduced species discovered in the Channel Islands in 2020 (Hamer *et al.* 2021).

–   Antenna with at least 10 segments   33

33   Mandible with a pointed tip but otherwise lacking distinct triangular teeth (B.37–B.38)   34

–   Mandible with a cutting edge bearing at least five triangular teeth (e.g. B.42, B.63)   35

34   Antenna with 10 or 11 segments; mandible short, with toothless cutting edge apart from the tooth at the tip (B.37)   *Tetramorium atratulum*

–   Antenna with 12 segments; mandible long, evenly curved like a scythe blade (B.38)
Strongylognathus testaceus

Tetramorium atratulum and *Strongylognathus testaceus* are both parasites in *Tetramorium caespitum* nests.

35   Antenna with 10 segments, the two at the tip forming a club (B.39); propodeum unarmed, without a pair of teeth or spines   *Solenopsis fugax*

–   Antenna with 11 or 12 segments; propodeum armed with a pair of teeth or spines (e.g. B.46, B.56)   36

Several introduced *Monomorium* species can be found in heated buildings, including the notorious pharaoh ant, *M. pharaonis*; these ants would key to this couplet, but have 12 antennal segments and an unarmed propodeum.

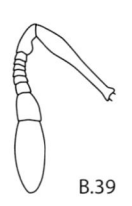

B.39

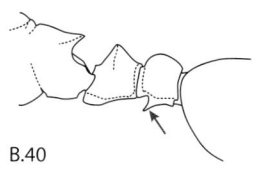

B.40

36   Postpetiole with a projecting spine below (B.40); upper surfaces of head and mesosoma glossy and mirror-like   *Formicoxenus nitidulus*

–   Postpetiole without a spine below (e.g. B.46, B.52); upper surfaces of head and mesosoma at least partly sculptured and often strongly wrinkled   37

37   Antenna with 11 segments   *Leptothorax acervorum*

–   Antenna with 12 segments   38

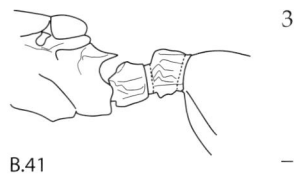

B.41

38   Lower surface of head with a ridge along each side (as A.50); petiole almost square or pentagonal in profile (B.41); front margin of clypeus with two prominent teeth, sometimes with an additional small one in the middle   *Myrmecina graminicola*

–   Lower surface of head without a pair of ridges; petiole not square, at least narrowing in front (e.g. B.46); front margin of clypeus without teeth   39

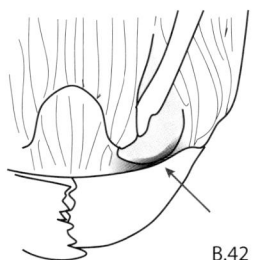

B.42

39   Sides of the clypeus raised into sharp semicircular ridges bordering the deep antennal sockets (B.42); mesoscutum and scutellum largely mirror-like with scattered fine punctures   *Tetramorium caespitum*

A closely related species, *Tetramorium impurum*, has been reported from the Channel Islands. If necessary, consult an expert. Several exotic species of *Tetramorium* are common in hothouses. In these species, the head has frontal ridges extending up to its rear corners.

–   Sides of the clypeus may be narrow but are not raised into sharp ridges, instead smoothly transitioning into front of clypeus (as in B.64); mesoscutum and scutellum at least finely wrinkled and usually strongly so   40

The clypeal character requires some experience. If in doubt, go with the mesosoma character.

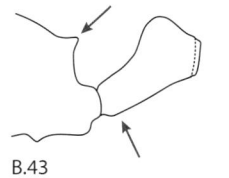

B.43

40   Petiole elongate in front and without a small tooth underneath (B.43)   41

–   Petiole narrowed in front but not elongate, and has a small tooth underneath (e.g. B.46–B.47)   43

41   Middle portion of clypeus with a pair of prominent ridges (as A.55–A.56)   genus *Stenamma* 42

–   Middle portion of clypeus without a pair of ridges   *Aphaenogaster subterranea*

Channel Islands only. In 2024, the presence of *Messor capitatus* was confirmed on Guernsey. It would key to this couplet, but is black rather than brown and has strongly polymorphic workers (pp. 33, 82).

42   Width of central smooth area of frons (frontal triangle) about one third of the total frons width (as A.55); viewed from above, front of petiole has distinct nubs on each side (as A.57)   *Stenamma debile*

–   Width of central smooth area of frons (frontal triangle) less than one third of the total frons width (as A.56); viewed from above, front of petiole lacks side nubs (as A.58)   *Stenamma westwoodii*

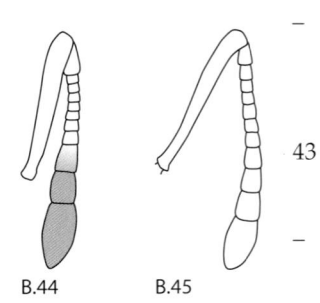

B.44   B.45

43   The last three antennal segments nearly as long as the remainder of the funiculus (B.44); mandible with 5 or 6 teeth   genus *Temnothorax* 44

–   The last three antennal segments distinctly shorter than the remainder of the funiculus (B.45); mandible occasionally with 6 but usually with 7 or more teeth   genus *Myrmica* 47

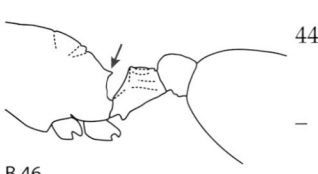

B.46

44   Antenna with terminal three segments the same pale brown colour as the remaining segments   *Temnothorax nylanderi*

–   Antenna with terminal three segments distinctly darker than the remaining segments (B.44)   45

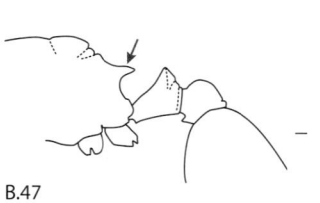

B.47

45   Petiole with peak forming an obtuse angle (B.46); scutellum with dense, delicate longitudinal furrows that are almost as pronounced as those on the mesoscutum; spines on propodeum short and triangular (B.46)   *Temnothorax albipennis*

–   Petiole with peak forming an acute angle (B.47); scutellum with weak longitudinal furrows that are much less pronounced than those on the mesoscutum; spines on propodeum may be long or short   46

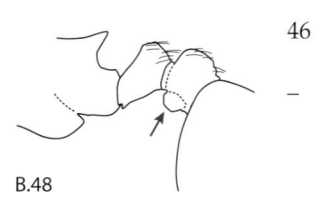

B.48

46   Mesosoma dark brown; spines on propodeum long and curved (B.47)   *Temnothorax interruptus*

–   Mesosoma yellow-brown; spines on propodeum short and triangular (even shorter than in B.46)   *Temnothorax unifasciatus*

Channel Islands only, with a single record on mainland Britain, presumably as an introduction.

47   Postpetiole with a projecting lobe below (B.48); head with a raised collar at the neck (B.49); first dorsal plate on gaster without standing hairs   *Myrmica karavajevi*

A rare social parasite without a worker caste, found in nests of *Myrmica sabuleti* and *M. scabrinodis*.

–   Postpetiole without a projection below (as B.56); head without modification at the neck; first dorsal plate on gaster with standing hairs   48

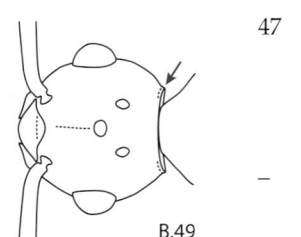

B.49

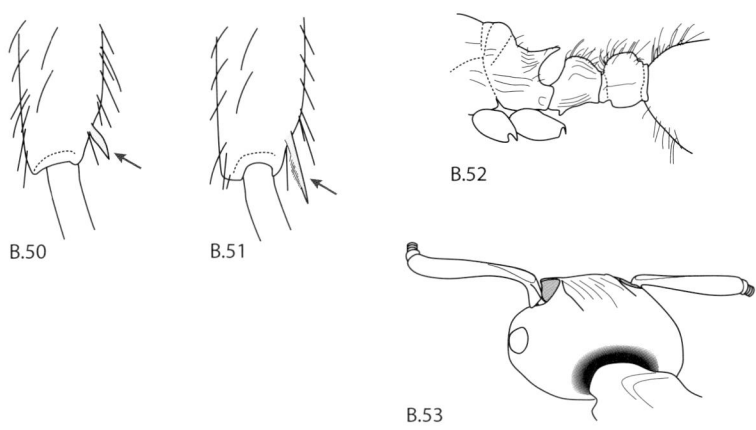

B.50

B.51

B.52

B.53

48 A small species, approximately 4.5 mm; mid tibia with very reduced spur (B.50); hairs of legs, antennae and body denser and more slanting, particularly noticeable on the antennal scapes; spines on propodeum shorter and thicker than in all the following *Myrmica* species except *M. rubra* (B.52)   *Myrmica hirsuta*

A rare species found only in the nests of its host *Myrmica sabuleti*.

– Larger species, usually 5.0 mm or more; mid tibia with comb-like spur (B.51); hairs less dense and more erect; spines on propodeum usually longer and narrower (e.g. B.57, B.68)   49

49 Viewed from the rear (as shown in B.53), base of antennal scape curved, sometimes strongly so; base of scape lacks ridges, flattened areas or outgrowths, and therefore remains approximately circular in cross-section throughout the curve (B.54–B.55)   50

– Viewed from the rear (as shown in B.53), base of antennal scape distinctly angled on the outside of an abrupt bend; base of scape with ridges, flattened areas or outgrowths (B.58–B.62)   52

Distinguishing *Myrmica* species relies heavily on the shape of the base of the antennal scape (i.e. towards its attachment to the head). The scape should be manipulated to extend laterally from the head. Rear view refers to looking from the rear of the head such that the funiculus points directly away from the observer, while front view refers to looking at the head face on. Photographs of front and rear views of the scape are shown in Fig. 6.9.

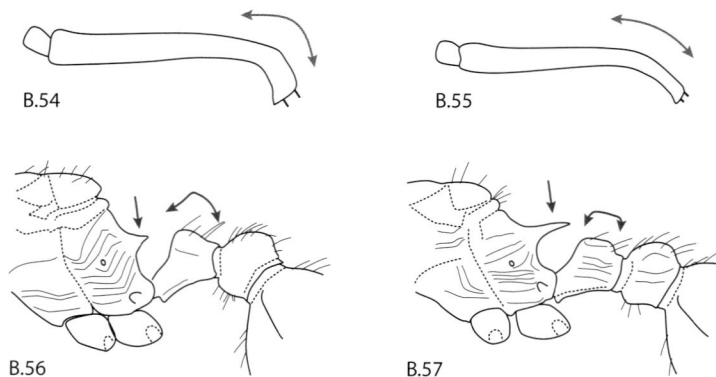

50   Scape thick and sharply curved near the base, nearly forming a right angle (B.54); frontal triangle extensively sculptured   *Myrmica sulcinodis*

–   Scape slender and gently curved in an obtuse angle (B.55); frontal triangle largely smooth, shining   51

51   Upper surface of petiole curved into the articulation with postpetiole (B.56); sides of petiole and postpetiole weakly wrinkled; spines on propodeum triangular, shorter than the distance between their tips

Myrmica rubra

A small form of queen has been distinguished as *Myrmica microrubra*, and may be a social parasite species.

–   Upper surface of petiole abruptly angled before meeting postpetiole (B.57); sides of petiole and postpetiole strongly wrinkled; spines on propodeum slender, about as long as the distance between their tips

Myrmica ruginodis

52   Bend of scape with an outgrowth projecting transversely across its outer corner, appearing tooth-like or flange-like when viewed from the rear (B.58–B.59)   53

–   Bend of scape without a transverse flange or tooth, but with a ridge or outgrowth running along the long axis of the scape (B.60–B.62)   54

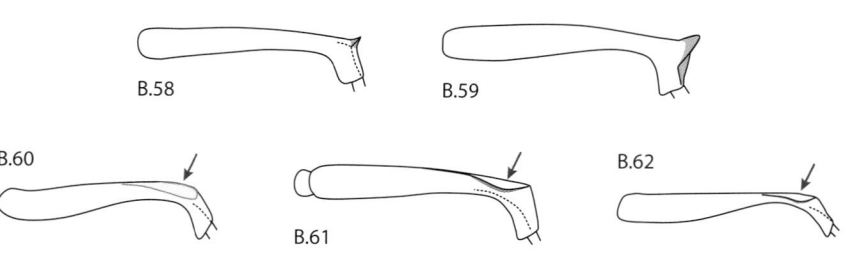

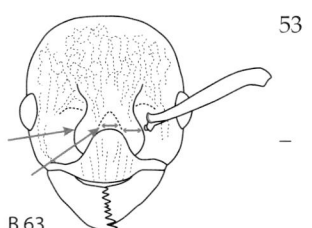

B.63

53   Outgrowth of scape bend is narrow and tooth-like (B.58); frontal lobes wide in comparison to the frontal triangle (B.63); body colour dark reddish-brown, sometimes with lighter mesosoma   *Myrmica lobicornis*

–   Outgrowth of scape broad and flange-like, wrapping round bend (B.59); frontal lobes narrow in comparison to the frontal triangle (B.64); body colour clear red
                                             Myrmica schencki

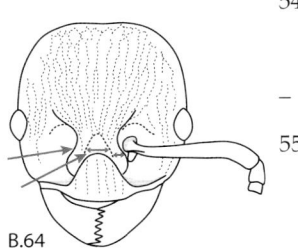

B.64

54   Clypeus with front margin slightly concave in middle   *Myrmica vandeli*

     Rare and little known in Britain

–   Clypeus with front margin straight or convex   55

55   Scape with a low ridge extending longitudinally from the bend. The ridge is inconspicuous but borders a flattened area of the surface that is angled such that it is visible not only from the front (B.65) but also when viewed strictly from the rear (B.53, B.60)
                                Myrmica scabrinodis

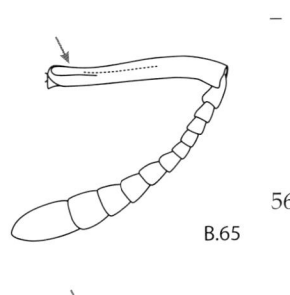

B.65

–   Scape with a sharp-edged projecting ridge extending longitudinally from the bend. The flattened surface of the projection faces the observer when viewed from the front (B.66–B.67), but is barely visible when viewed strictly from the rear (as its sharp edge projects towards the observer) (B.61–B.62)   56

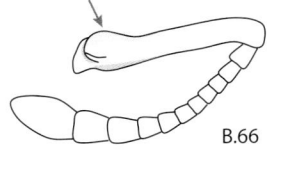

B.66

56   Petiole with an abrupt angle between upper margin and rear margin (B.68); outgrowth of scape usually robust (B.66)   *Myrmica sabuleti*

     A montane form of *M. sabuleti* with a larger, upswept antennal outgrowth has been considered a separate species, *Myrmica lonae*, although DNA analysis has cast doubt on this. This form has been found in Scotland.

–   Petiole with upper margin evenly curved to its join with the postpetiole (B.69); outgrowth of scape sharp but usually low (B.67)   *Myrmica specioides*

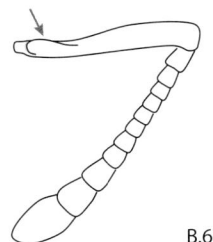

B.67   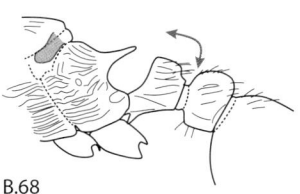 B.68

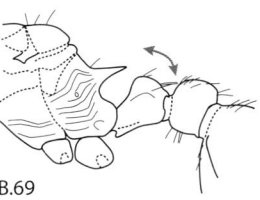 B.69

Key C: Males

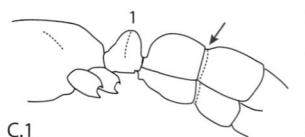

C.1

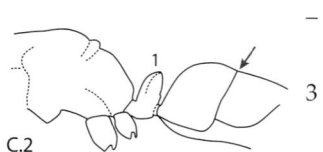

C.2

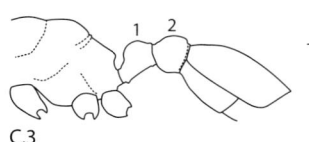

C.3

1   Waist consisting of 1 segment, the petiole (C.1–C.2) 2

–   Waist consisting of 2 segments, the petiole and postpetiole (C.3)   Myrmicinae 30

2   Gaster with a distinct constriction between the first and second segments (C.1)   Ponerinae 3

–   Gaster with no constriction between the first and second segments (C.2)   4

3   Winged; plate on underside of last segment of gaster elongated into a down-curved sting-like spine
                    Ponera coarctata and *Ponera testacea*

There are no established characters to separate males of these two species.

–   Wingless and worker-like; plate on underside of last segment of gaster not elongated
                                    Hypoponera punctatissima

This species has occasionally been found outdoors. It and a related species, *Hypoponera ergatandria*, are more often encountered in hothouses etc.; they are only separable by precise quantitative measurements – if necessary, consult an expert.

4   No erect hairs on top of mesosoma (C.4)
                                    Dolichoderinae 5

–   Few to many erect hairs present on top of mesosoma (e.g. C.5)   Formicinae 7

5   Front margin of clypeus straight to broadly convex; antenna with scape short, not reaching hind margin of head   *Linepithema humile*

This is the infamous invasive Argentine ant; it has very occasionally been found outdoors in urban areas of Britain. A related species, *Linepithema iniquum*, has been found several times in hothouses.

–   Front margin of clypeus with a shallow notch in centre; antenna with scape long, extending well beyond hind margin of head   genus *Tapinoma* 6

C.4

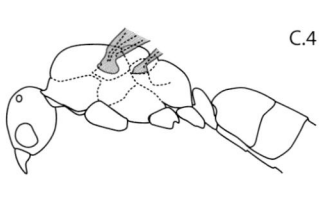

C.5

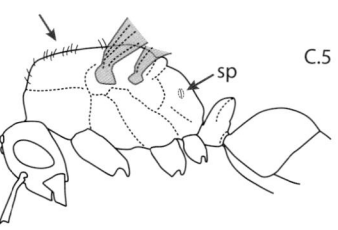

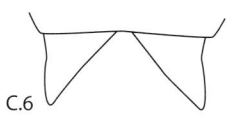

C.6

C.7

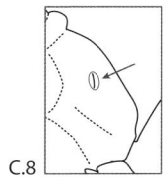

C.8

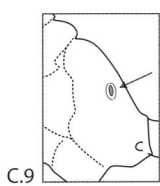

C.9

6   Head rounded behind; side lobes of subgenital plate (plate on underside of last segment of gaster) diverging and narrow (C.6)   *Tapinoma subboreale*

–   Head squarer behind; side lobes of subgenital plate closer together and broader (C.7)   *Tapinoma erraticum*

7   Antenna with 12 segments; a tiny species measuring < 2.0 mm   *Plagiolepis pallescens*

Channel Islands only (although there are occasional sporadic reports of *Plagiolepis* species on the south coast). An exotic species, *Plagiolepis alluaudi*, is frequently found in hothouses of botanical gardens.

–   Antenna with 13 segments; larger species, at least 2.5 mm and usually longer   8

8   Opening of propodeal spiracle elongate, slit-like, on side of propodeum (C.5 sp, C.8); ant generally larger (>7 mm)   genus *Formica* 9

–   Opening of propodeal spiracle circular or nearly so, on curved surface where it rounds into hind face (C.9); ant generally smaller (at most 5 mm)   genus *Lasius* 19

Be sure to observe the shape of the opening rather than the surrounding rim.

9   Front margin of clypeus with a notch in the middle (C.10)   *Formica sanguinea*

–   Front margin of clypeus smoothly rounded (C.11– C.12)   10

10   Hind margin of head strongly concave (C.11)   *Formica exsecta*

–   Hind margin of head not strongly concave (C.12)   11

11   Compound eye with minute erect hairs between the facets (use high magnification and vary the lighting)   12

–   Compound eye without erect hairs   15

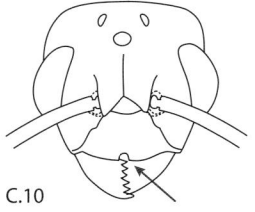

C.10

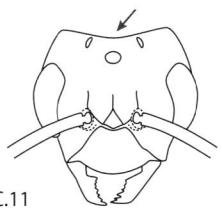

C.11

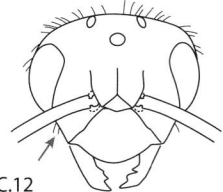

C.12

12 Side of head between compound eye and clypeus without long erect hairs *Formica rufa*

Research suggests that the species known in Britain as *F. rufa* is in fact a stable hybrid of two non-British species, *Formica polyctena* and *F. rufa* (Seifert *et al.* 2010).

– Side of head between compound eye and clypeus with few-to-many long erect hairs (C.12) 13

13 Gaster dull, not shining; erect hairs arising over entire upper surface of second segment of gaster *Formica pratensis*

Channel Islands only; extinct in mainland Britain.

– Gaster moderately shining; erect hairs arising mainly from front portion of upper surface of second segment of gaster 14

14 Side of head between eye and clypeus with only two or three erect hairs (C.12); gaster with only a few erect hairs *Formica aquilonia*

– Side of head between eye and clypeus with numerous erect hairs; gaster with numerous scattered erect hairs *Formica lugubris*

15 Underside of head with one or two long hairs; gaster glossy *Formica picea*

– Underside of head without hairs; gaster dull or diffusely shining 16

16 Upper margin of petiole scale with fringe of very short hairs only *Formica fusca*

– Upper margin of petiole scale with fringe of short hairs and at least some longer erect hairs (C.13) 17

17 Scutellum more shining than mesoscutum; gaster diffusely shining *Formica lemani*

– Scutellum and mesoscutum both equally dull; gaster silky and rather matt 18

18 Femur partly dark *Formica rufibarbis*

Rare and highly endangered on mainland Britain. Also well known from the Isles of Scilly.

– Femur uniformly pale *Formica cunicularia*

19 Usually glossy jet black; opening of metapleural gland lacks collar of guard hairs (like B.22 mpg) *Lasius fuliginosus*

– At most brownish-black; opening of metapleural gland surrounded by guard hair collar (like B.25 mpg) 20

C.13

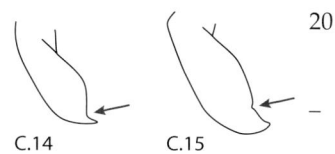

C.14   C.15

20   Mandible with one tooth at the tip, followed by a smoothly curved cutting margin with at most a slight notch (C.14–C.15)   21

–   Mandible with a tooth at the tip, followed by a pronounced notch and/or further teeth (C.20–C.21, C.23, C.25–C.27)   25

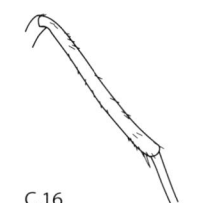

C.16

21   Tibia with short slanting hairs arising among the pubescence (C.16)   22

–   Tibia pubescent only (as defined in Fig. 6.7)   24

22   Frons and frontal triangle polished; mesosoma brown to brownish-black   23

–   Frons and frontal triangle dull; mesosoma distinctly reddish behind the wing bases   *Lasius emarginatus*

Native to Channel Islands and recently established in southern England.

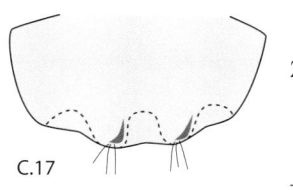

C.17

23   Subgenital plate (plate on underside of last segment of gaster) with central area bearing two embossed areas, but rear margin only weakly projecting (C.17)
          Lasius niger

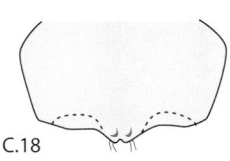

C.18

–   Subgenital plate with central area nearly flat or with small embossed areas, and the rear margin more strongly projecting in the middle (C.18)
          Lasius platythorax

These males are not easy to separate, dissection of the subgenital plate may help.

24   Small, 2–3 mm long; head without projecting hairs in the region of the ocelli   *Lasius neglectus*

Invasive species; males and winged queens rarely if ever leave the nest.

–   Larger, at least 3.5 mm long; head with projecting hairs in the region of the ocelli
          Lasius alienus and *Lasius psammophilus*

Males participate in mating flights as normal. Males of these two species are not readily separable.

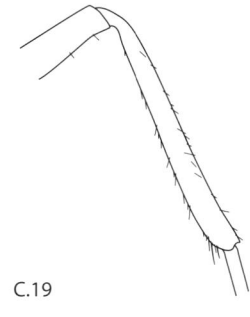

C.19

25   Hind tibia with short, fine hairs projecting above the pubescence (C.19)   26

–   Hind tibia with pubescence only   27

The projecting hairs can be hard to see, backlighting may help.

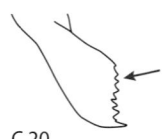

C.20

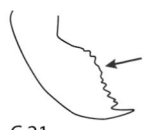

C.21

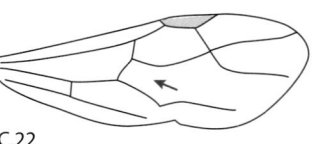

C.22

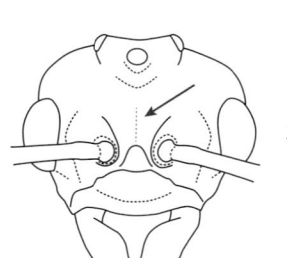

C.23

26   Frons shining, with very little microsculpture; mandible with pronounced teeth (C.20)   *Lasius meridionalis*

–   Frons dull, microsculptured; mandible with teeth less distinct (C.21)   *Lasius umbratus*

These species are difficult to separate with certainty due to overlapping characters.

27   Discoidal cell (area enclosed by veins) usually absent from the middle of the wing (C.22); frontal line of head indistinct (C.23)   *Lasius flavus*

A closely related species, *Lasius myops*, has been reported (rarely) in the Channel Islands and possibly mainland Britain. The male is similar to *L. flavus*, but usually has more teeth on its mandible (compared to C.23).

–   Discoidal cell usually present in the middle of the wing (C.24); frontal line distinct, reaching close to the middle ocellus (C.25)   28

28   Sides of head behind compound eyes usually without hairs; wing smoky near base; mandible with a tooth at the tip followed by a deep notch (C.25, C.26)   *Lasius brunneus*

–   Sides of head behind compound eyes with hairs; wing clear; mandible with at least two teeth (C.27)   29

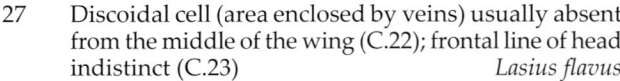

C.24

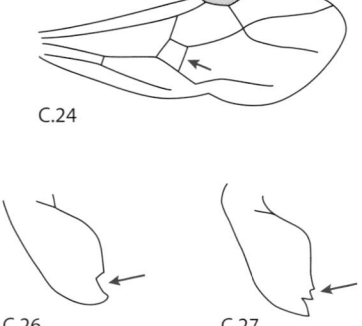

C.26   C.27

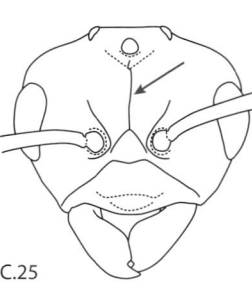

C.25

29   Mandible with two teeth, one at the tip and one just before the tip, followed by 0–2 more tiny teeth (C.27) *Lasius mixtus*

–   Mandible with two teeth, one at the tip and one just before the tip, followed by 2–6 more tiny teeth
Lasius sabularum

These species are difficult to separate due to overlapping characters.

30   Wingless   31

–   Winged   32

31   Antenna with 10 or 11 segments; body surface dull and poorly pigmented; mesosoma enlarged, reminiscent of the mesosoma of winged ants
Tetramorium atratulum

–   Antenna with 12 segments; body surface strongly shining, orange to dark brown; mesosoma reduced so that ant is worker-like in appearance
Formicoxenus nitidulus

32   Antenna with 10 segments, with third segment elongate (C.28)   33

–   Antenna with 12 or 13 segments, with third segment not outstandingly elongate   34

33   Mandible slender, curved, with no teeth (C.29); node of petiole surmounted by two teeth
Strongylognathus testaceus

–   Mandible with teeth on a broad cutting edge (C.30); node of petiole more or less rectangular
Tetramorium caespitum

A closely related species, *Tetramorium impurum*, has been reported from the Channel Islands. See Wagner *et al.* (2017) for characters to distinguish these species, and if necessary, consult an expert.

34   Antenna with 12 segments   35

–   Antenna with 13 segments   37

The males of *Strumigenys perplexa*, discovered in the Channel Islands in 2020, are poorly known, but would probably key out here as a small winged myrmicine ant with 13 antennal segments and an extremely short scape.

35   Propodeum curved into petiole, without spines or teeth; mesoscutum lacking a pair of diagonal grooves (notauli) *Solenopsis fugax*

–   Propodeum with a pair of triangular teeth; mesoscutum with notauli (as in C.31)   36

C.28

C.29

C.30

C.31

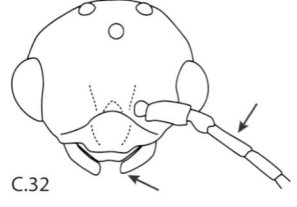

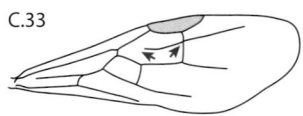

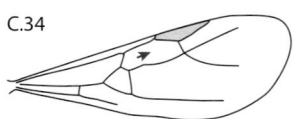

C.32

C.33

C.34

36   Mandibles very small and truncate, not capable of closing to touch each other (C.32); abundant hairs on mesosoma; antenna with scape short and segments 3–12 elongate (C.32)   *Leptothorax acervorum*

–   Mandibles fully formed with distinct teeth, capable of closing; mesosoma almost without hairs; antenna with scape longer and segments 3–12 short (like C.40)
    Myrmica karavajevi

A rare social parasite without a worker caste, found in nests of *Myrmica sabuleti* and *M. scabrinodis*.

37   Forewing with a vein that is incomplete, thereby forming a cross-like pattern (C.33); in most species, spurs of hind tibia finely toothed, comb-like (as in C.39)   genus *Myrmica* (part) 38

–   Forewing with veins complete so that cells do not have a vein extending into them (C.34); spurs of hind tibia not comb-like or are absent   47

38   A small species with an unusually broad postpetiole, broader than high; hairs on petiole and postpetiole clearly longer than width of hind femur
    Myrmica hirsuta

A rare species found only in the nests of its host *Myrmica sabuleti*.

–   Larger species with postpetiole generally of normal breadth, higher than broad; hairs on petiole and postpetiole about as long as width of hind femur   39

39   Mesoscutum immediately in front of notauli wrinkled, obscuring the notauli; petiole strongly sculptured
    Myrmica sulcinodis

–   Mesoscutum in front of notauli smooth and usually shining with notauli very distinct (as in Fig. 6.4); petiole weakly sculptured   40

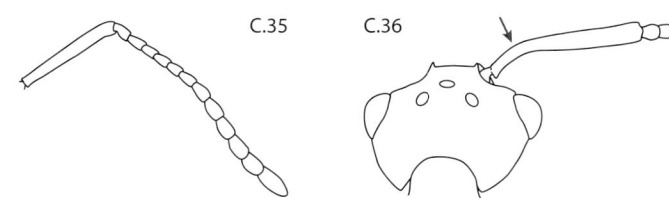

C.35   C.36

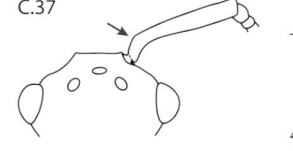

C.37

40 Antennal scape long and slender, at least as long as the basal six segments of the funiculus (C.35), with a shallow curve at base when viewed from rear (C.36)   41

–   Antennal scape shorter than the basal six segments of the funiculus (C.40–C.43), if approaching this length then it has an abrupt bend at base (C.37)   42

41 Frons largely smooth with only a hint of longitudinal lines; hind tibia with short hairs, not projecting (C.38)   *Myrmica ruginodis*

–   Frons roughened with very fine longitudinal lines; hind tibia with some longer, projecting hairs (C.39)   *Myrmica rubra*

42 Antennal scape longer than the basal four segments of funiculus (C.40–C.41)   43

–   Antennal scape shorter than the basal four segments of funiculus (C.42–C.43)   45

43 Scape longer than basal five segments of funiculus (C.40), sharply bent at base (C.37)   *Myrmica lobicornis*

–   Scape thicker, longer than four but shorter than five basal segments of funiculus, only weakly bent (C.41)   44

44 Front edge of clypeus flat or convex   *Myrmica sabuleti*

–   Front edge of clypeus slightly concave in middle   *Myrmica vandeli*

Rare and little known in Britain.

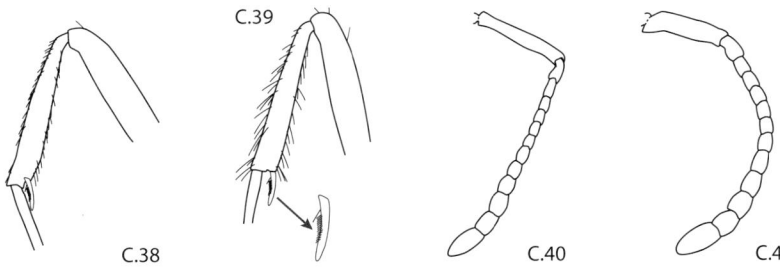

C.38   C.39   C.40   C.41

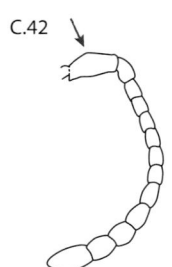

C.42

45   Scape thick, somewhat humped in the middle (C.42); hind tarsus with numerous long, strongly projecting hairs   *Myrmica scabrinodis*

–   Scape more evenly cylindrical (C.43); hind tarsus with hairs short and not strongly projecting   46

46   Frons with circular impression just forward of middle ocellus; scape angled at base (C.44)   *Myrmica schencki*

–   Frons not impressed forward of middle ocellus; scape barely angled at base   *Myrmica specioides*

C.43

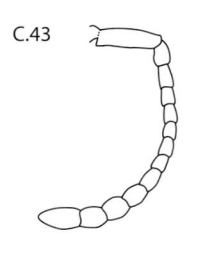

47   Mandibles reduced to tiny stubs, not functional (C.45); wings smoky   *Myrmecina graminicola*

–   Mandibles of normal size and toothed; wings clear   48

48   Forewing with 2 cells along the middle (C.46)

Aphaenogaster subterranea

Channel Islands only. In 2024, the presence of *Messor capitatus* was confirmed on Guernsey. It would key to this couplet, but is black rather than brown and has strongly polymorphic workers (pp. 33, 82).

–   Forewing with 0–1 cell along the middle (as in C.34)   49

49   Petiole elongate in front (C.47); propodeum with a pair of small triangular teeth   genus *Stenamma* 50

–   Petiole narrowed in front but not elongate (C.48); propodeum angled or rounded, but without small teeth   genus *Temnothorax* 51

C.44

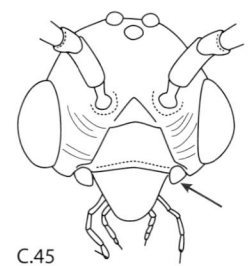

C.45

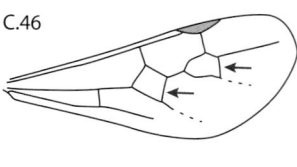

C.46

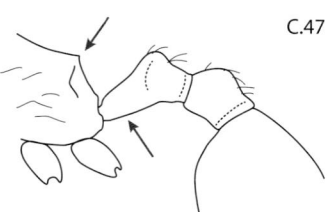

C.47

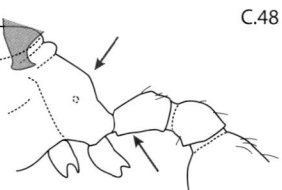

C.48

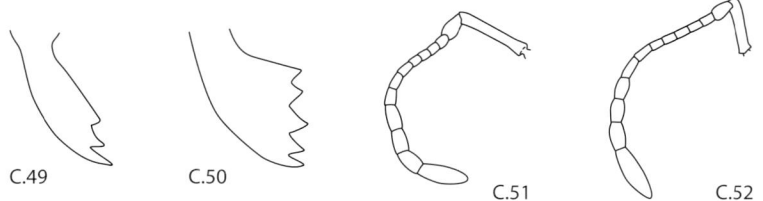

C.49 C.50 C.51 C.52

50   Mandible with 3 teeth, appearing narrow with no
      marked cutting edge (C.49)   *Stenamma debile*

–   Mandible with 5 teeth on a broad cutting edge
      (C.50)   *Stenamma westwoodii*

51   Front portion of mesoscutum between the notauli
      (Fig. 6.4) smooth and glossy *Temnothorax nylanderi*

–   Front portion of mesoscutum between the notauli
      weakly to strongly sculptured even if shining   52

52   Scutellum sculptured on top apart from a narrow
      smooth strip in the middle; antennal segments 3–5
      as broad as long (C.51)   *Temnothorax interruptus*

–   Scutellum smooth on top; antennal segments 3–5
      longer than broad (C.52)   53

53   Front portion of mesoscutum between notauli strongly
      wrinkled; legs and antennae light brown
                                    Temnothorax albipennis

–   Front portion of mesoscutum between notauli faintly
      sculptured; legs and antennae colourless
                                    Temnothorax unifasciatus

      Channel Islands only with a single record on mainland
      Britain, presumably as an introduction.

Key D: Quick-check field key to workers of common and distinctive British and Irish ants

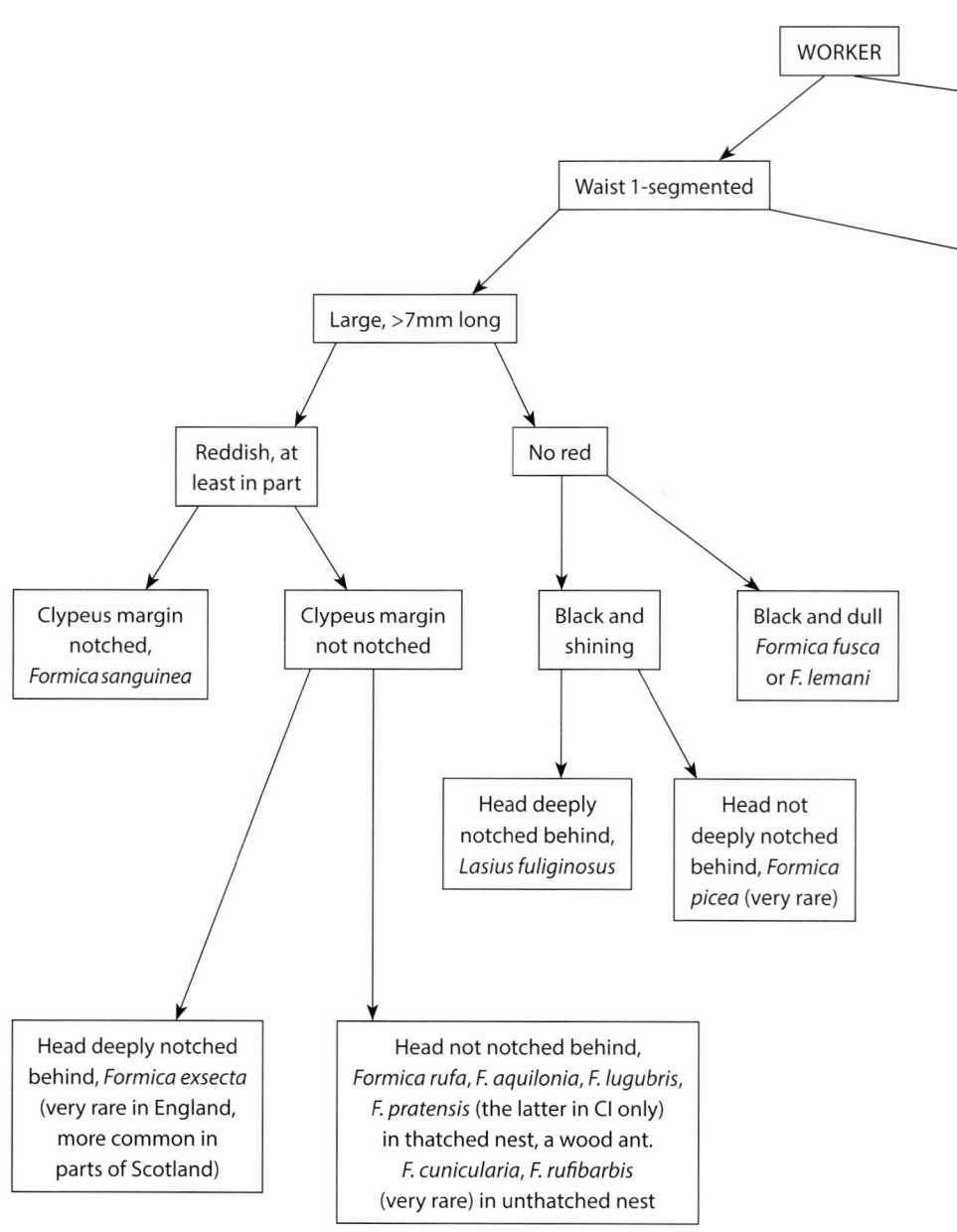

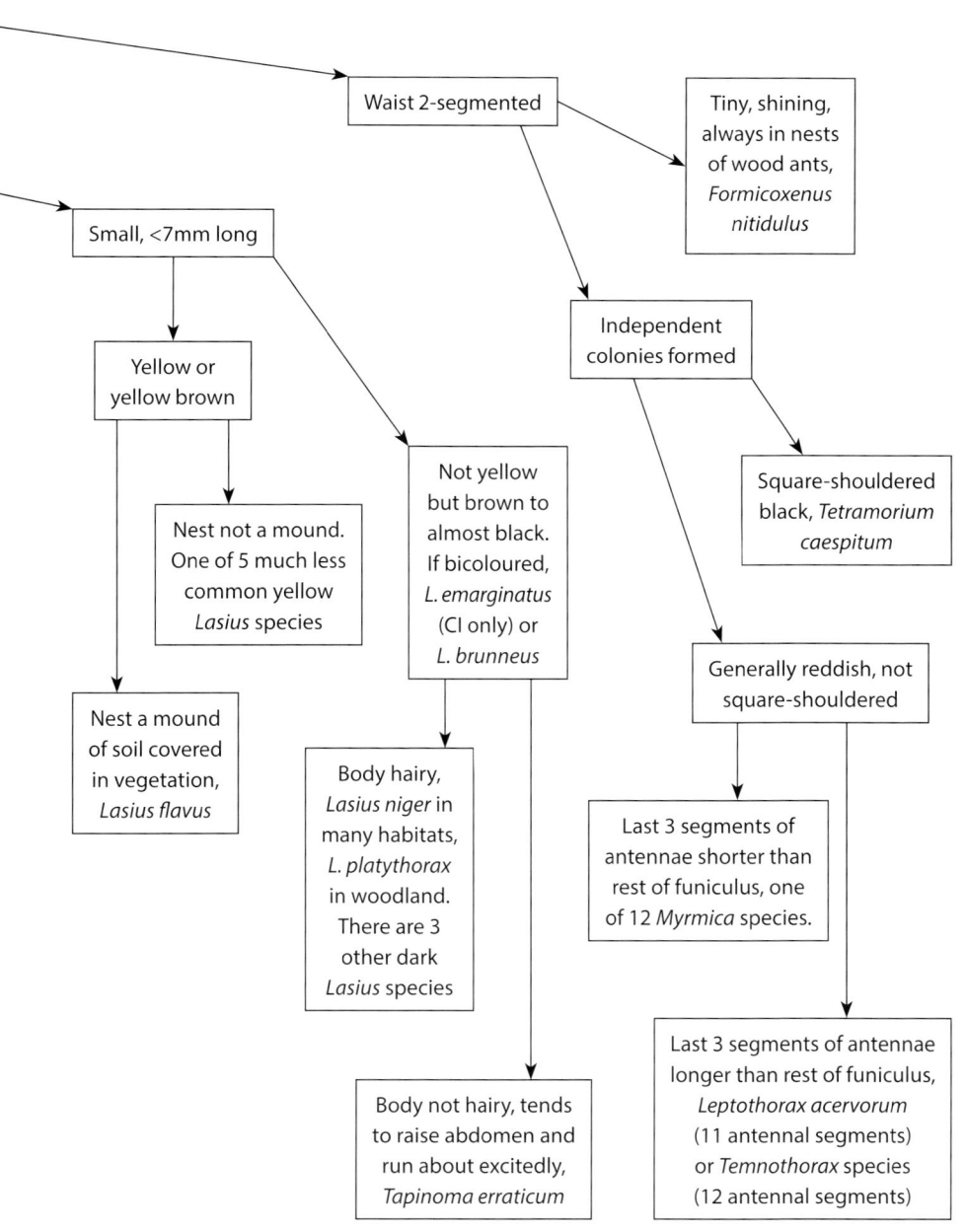

Key E: Quick-check key to nests of common and distinctive species of British and Irish ants

Many of our ants make nests in the soil, under logs and stones. Some are much more distinctive, however. This key to nests should be used in conjunction with the quick-check field key and/or the main keys.

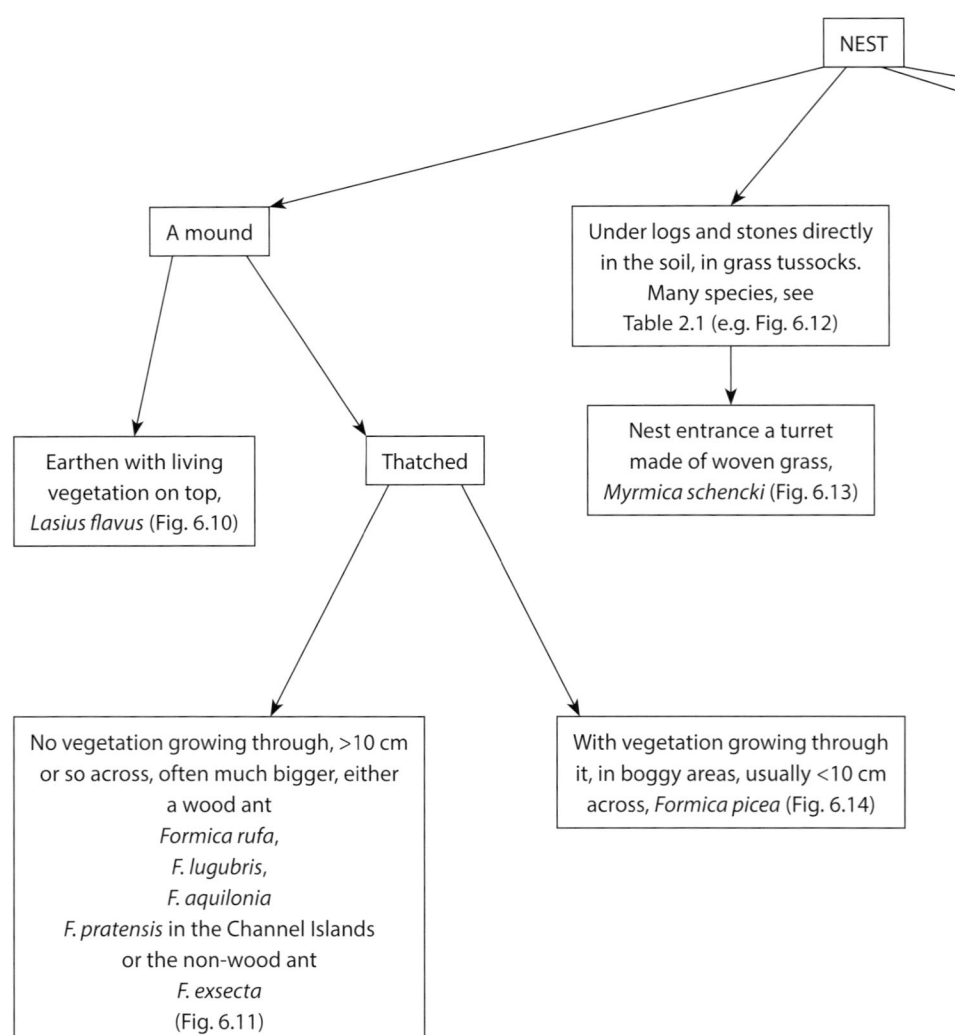

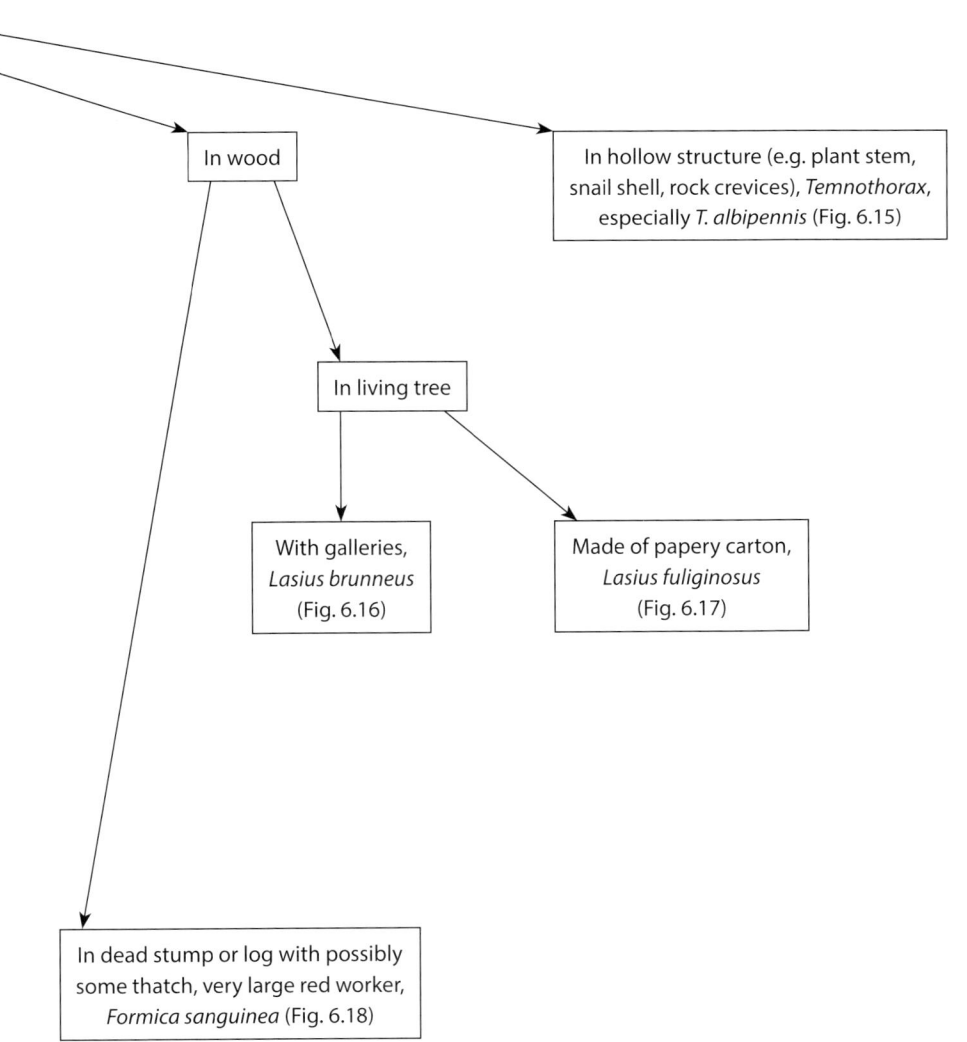

Fig. 6.10 Nest mounds of *Lasius flavus*

Fig. 6.11 Mound of *Formica aquilonia*

Fig. 6.12 *Lasius flavus* nesting under a stone

Fig. 6.13 Woven turret of *Myrmica schencki* nest with guard worker (Rosemary Winnall)

Fig. 6.14 Thatched mound of *Formica picea*

Fig. 6.15 *Temnothorax albipennis* in a hollow, dead stem of viper's bugloss *Echium vulgare*

Fig. 6.16 *Lasius brunneus* nest in tree trunk (Peter Furze)

Fig. 6.17 Carton nest of *Lasius fuliginosus* (Grzegorz Wagner)

Fig. 6.18 Loose mound nest of *Formica sanguinea* in dead tree stump

6.4 Old names and common names

taxonomy
The classification and
naming of things.

Research never stops. This is true in all areas of science, including taxonomy. During the almost three decades since the publication of the first edition of this book there have been several name changes. For example, in 1996 there were five species of *Leptothorax* ants in Britain and one in Ireland. However, in the intervening years, four of these have been reclassified as *Temnothorax*, leaving only one member of this genus in the region. Older references will discuss these ants by these old names, so it is quite important to be aware of the major changes. Table 6.1 shows some important name changes that have occurred over the last 50 years or so. As noted earlier (p. 228), ants are all rather similar in appearance and this has rather discouraged the use of common names. However, some do have these. Wragge Morley (1953) listed all the species covered by his book under a common name. The widely accepted common names, and Wragge Morley's names, are given in Table 6.1.

Table 6.1 Current and former Latin names of ant species, together with some of the common names that have been used.

Current Latin name (2023)	Some older Latin names	Common name (WM: names in Wragge Morley 1953)
Aphaenogaster subterranea	n/a	n/a
Formica aquilonia	*Formica rufa*	Scottish wood ant
Formica cunicularia	*Formica fusca* var. *rubescens*	n/a
Formica exsecta	n/a	narrow-headed ant (rarer wood ant and rover wood ant in WM)
Formica fusca	n/a	dusky ant, silky ant (large black ant in WM)
Formica lemani	*Formica fusca*	as *F. fusca*
Formica lugubris	*Formica rufa*	hairy wood ant, northern wood ant
Formica picea	*Formica transkaucasica, F. candida*	bog ant (black bog ant in WM)
Formica pratensis	*F. nigricans*	meadow wood ant, black-backed meadow ant
Formica rufa	n/a	southern wood ant (wood ant in WM)
Formica rufibarbis	n/a	red-barbed ant (ruddy black ant in WM)
Formica sanguinea	n/a	blood red ant, red robber ant (blood-red slave maker in WM)
Formicoxenus nitidulus	n/a	shining guest ant (rarer guest ant in WM)
Hypoponera ergatandria	*Hypoponera schauinslandi*	n/a
Hypoponera punctatissima	n/a	n/a
Lasius alienus	*Donisthorpea aliena*	n/a
Lasius brunneus	*Acanthomyops brunneus*	(brown tree ant in WM)
Lasius emarginatus	n/a	n/a
Lasius flavus	*Acanthomyops flavus*	yellow meadow ant, yellow hill ant
Lasius fuliginosus	*Donisthorpea fuliginosa, Acanthomyops fuliginosus*	jet ant (jet black ant in WM)

(cont.)

Current Latin name (2023)	Some older Latin names	Common name (WM: names in Wragge Morley 1953)
Lasius meridionalis	n/a	n/a
Lasius mixtus	n/a	n/a
Lasius myops	n/a	n/a
Lasius neglectus	n/a	n/a
Lasius niger	*Acanthomyops niger, Donisthorpea nigra*	black garden ant, common black ant (black lawn ant in WM)
Lasius platythorax	n/a	n/a
Lasius psammophilus	n/a	n/a
Lasius sabularum	n/a	n/a
Lasius umbratus	*Donisthorpea umbrata, Acanthomyops umbratus*	(yellow lawn ant in WM)
Leptothorax acervorum	n/a	(common guest ant in WM)
Linepithema humile	*Iridomyrmex humilis*	Argentine ant
Monomorium pharaonis	n/a	pharaoh ant (pharaoh's ant in WM)
Myrmecina graminicola	n/a	(woodlouse ant in WM)
Myrmica hirsuta	n/a	guest ant
Myrmica karavajevi	*Sifolinia karavajevi*	guest ant
Myrmica lobicornis	n/a	n/a
Myrmica lonae	n/a	n/a
Myrmica rubra	*Myrmica laevinodis*	European fire ant (erroneous name), common red ant
Myrmica ruginodis	n/a	n/a
Myrmica sabuleti	n/a	n/a
Myrmica scabrinodis		elbowed red ant
Myrmica schencki	n/a	n/a
Myrmica specioides	*Myrmica bessarabica, Myrmica puerilis*	n/a
Myrmica sulcinodis		elegant red ant
Myrmica vandeli	n/a	n/a
Plagiolepis pallescens	*P. taurica, P. vindobonensis*	n/a
Ponera coarctata		fossil ant, indolent ant
Ponera testacea	*Ponera coarctata*	n/a
Solenopsis fugax		thief ant
Stenamma debile	*Stenamma westwoodii*	n/a
Stenamma westwoodii		Westwood's ant
Strongylognathus testaceus		testaceous guest ant
Strumigenys perplexa	n/a	n/a
Tapinoma erraticum	n/a	erratic ant
Tapinoma subboreale	*Tapinoma erraticum, Tapinoma ambiguum*	n/a
Temnothorax albipennis	*Leptothorax tuberum, Leptothorax tuberointerruptus*	acorn ant
Temnothorax interruptus	*Leptothorax interruptus*	long-spined ant, slender-bodied ant
Temnothorax nylanderi	*Leptothorax nylanderi*	acorn ant
Temnothorax unifasciatus	*Leptothorax unifasciatus*	acorn ant
Tetramorium atratulum	*Anergates atratulus*	dark guest ant (workerless ant in WM)
Tetramorium caespitum		square-shouldered ant

7 Studying ants

7.1 Recording and collecting

voucher specimen

A preserved specimen of an organism that acts as a permanent record.

A fundamental stage in our understanding of any living thing is to know where it lives. Ants, despite their familiarity, are often poorly known. This is in part due to identification difficulties and in part because of their small size and secretive habits. Thus, the amateur can make a very useful contribution to our knowledge. The BWARS website lists the species, shown in Table 7.1, that are under-recorded in some way.

Targeting these species in likely habitats where they are currently not recorded would be a very useful strategy. In addition, the targeting of grid squares that have never had any ant records could be productive too. These can be seen in Fig. 4.2.

Records of potentially interesting finds can be submitted to BWARS (see p. 303). It should be noted that, at least for some of the more difficult taxa, verification should be sought. For this reason, voucher specimens should be taken, retained and labelled with all relevant data (where and when collected, together with collector name as a minimum).

Ants can be encountered almost everywhere, even in urban areas and gardens. Many species forage on the ground and, if the weather is not too cold, they can usually be found by watching an area of ground closely for a few minutes. Another favoured technique for the collection of such species is the pitfall trap. This consists of a glass jar or plastic cup, sunk into a hole in the ground to its rim (Fig. 7.1).

Fig. 7.1 A pitfall trap

Surface-running species will fall in. If live specimens are needed, the trap should be used dry with the inner surfaces smeared with a suspension of PTFE (polytetrafluoroethylene) to prevent ants from escaping, and it should be checked frequently. If dead, preserved specimens will suffice, escapes can be prevented by charging the trap to a depth of about 2 cm with a preservative such as 70% isopropanol or antifreeze with a drop of detergent (washing-up liquid) to reduce surface tension. If the trap can be checked every few hours, preservative need not be added. Specimens caught in this way will not be so good for the museum drawer as those caught alive and killed in one of the ways described below. When left in the liquid the dead ant's body swells and becomes distorted, and when dried it is rigid. The specimen can be partially relaxed by leaving it for about 24 hours in a humid atmosphere in a container with a tight-fitting lid. In the bottom should be placed some thymol crystals and then absorbent material such as paper tissue or cellulose wadding moistened with water. The water must not run onto the specimen, so the lid is lined with blotting paper to absorb any moisture that condenses there. Alternatively, chopped cherry-laurel leaves can be used as a relaxing agent as they provide moisture and give off hydrogen cyanide which inhibits mould growth, but these must be treated with great care because hydrogen cyanide is very poisonous. Some of the firms mentioned on pp. 301–302 supply a relaxing fluid that can be painted directly onto stiff insects. Freshly killed specimens are always to be preferred for making up a serious collection (see below). Ants must be handled carefully to

Fig. 7.2 A pooter

Fig. 7.3 A sweep net

avoid damage. Stiff, narrow forceps are usually adequate, but fine watchmaker's forceps are best for very small specimens (Hölldobler & Wilson 1990).

Alternatively, specimens can be sucked up into a pooter (aspirator) (Fig. 7.2). This method is inconvenient for handling some of the larger wood ants, which produce significant quantities of formic (methanoic) acid, causing the collector discomfort when sucking at the pooter. The discomfort can be reduced by lining the bottom of the pooter tube with absorbent material such as plaster of Paris or, in the field in an emergency, with a little damp soil.

To collect ants from herbaceous vegetation and low branches, a sweep net is very effective. This consists of a long, stout calico bag supported on a (usually) D-shaped metal frame (Fig. 7.3). It is dragged, or 'swept', vigorously through the vegetation and the end is flicked over the rim to close it. The contents are then inspected carefully, while having a pooter, specimen tubes and forceps handy so that specimens can be transferred quickly. To sample more widely in a tree, beating may be used (Fig. 7.4). A large, pale sheet is laid out below selected branches and then these are struck firmly with a stick or shaken gently. Many crawling insects

Fig. 7.4 Net curtain being used as a beating tray

are dislodged and fall onto the sheet, from which they can be picked up easily with forceps, a pooter or a specimen tube.

These techniques work well for many species of ant in Britain, but it is harder to find the small species that nest in rotting logs and twigs and do not go far afield to forage, those living mainly underground and those inhabiting the nests of other species. In such cases, it is often worth searching for the nest rather than the ants themselves. For example, nests are usually the first clue to the presence of the yellow ant *Lasius flavus*. These ants forage mainly underground, and workers are rarely seen at the surface, but the nests are very conspicuous (e.g. p. 274). Another good way of searching out nests of terrestrial species is to look underneath stones and logs lying on the ground. Workers, males, queens and immature stages can often be sampled from such sites. The stones or logs should be replaced carefully, in exactly the original position, as soon as possible. Some species nest in rotting logs on the ground and the rotten wood is worth searching, preferably in a white tray. In the same way, a white, or at least pale, dish can be used to catch organisms from sieved leaf litter. This is a very productive method for species such as the two *Stenamma* species, which do not forage openly (Fig. 7.5). The exotic species *Strumigenys perplexa* has recently been discovered living in the wild in Guernsey using this method (Figs 7.6, 7.7).

Methods used for the capture of other flying insects, such as moths, can be used to catch the alates of ants. Thus, these are often found in moth traps at the right time of year, as well as being caught in Malaise traps (Figs 7.8, 7.9).

alate ant
An ant with wings, which are males, or queens before they shed their wings, at which point they become dealates.

Fig. 7.5 Sieving leaf litter

Fig. 7.6 A *Strumigenys perplexa* worker sieved from leaf litter by a roadside

Fig. 7.7 *Strumigenys perplexa* having caught a springtail (*Entomobrya intermedia*, a collembolan) (Andy Marquis)

Fig. 7.8 A mains-powered Heath-type moth trap. Versions that run off batteries with LED lights are now available. Other moth trap designs would also be suitable.

Fig. 7.9 A Malaise trap for flying insects (NHBS Ltd)

Some ants will come to bait. The case of *Myrmecina graminicola* has been mentioned (Fig. 7.10). Oddly, *Myrmica schencki* seems to favour cake crumbs (Fox, personal communication 2021) (Fig. 7.11). Some, such as *Ponera* species, are drawn to raw minced meat, and many species will gather around sugar solutions.

Another way to attract ants is to lay out tiles or flat stones to provide suitable nest sites. This needs to be done at the

Fig. 7.10 *Myrmecina graminicola* workers attracted to an orange

Fig. 7.11 *Myrmica schencki* workers attracted to (pink) cake crumbs (Mike Fox)

beginning of the season and checked throughout the spring and summer (Fig. 7.12).

A combination of the methods outlined above, applied with some care in a particular habitat, should give a complete species list.

Fig. 7.12 A tile with nest galleries of *Tetramorium caespitum* underneath

7.2 Preserving

A reference collection of ants is sometimes required so that identifications can be checked if needed. Specimens are best killed in ethyl acetate vapour, but care must be taken not to inhale this or get it on the skin. A glass specimen tube or screw-top jar should be used, because ethyl acetate attacks many plastics. A few drops of ethyl acetate are absorbed onto plaster of Paris or balls of tissue overlain with more tissue. The ants should be left in this killing bottle for an hour or so to ensure that they are quite dead. Ethyl acetate leaves the dead insects in a relaxed state, ideal for mounting neatly. They are best mounted straight away on a piece of stiff white card, usually 300 gsm Bristol board, big enough to hold the insect with its legs (and wings if present) fully spread out, leaving a small margin all around the edge with extra space for the pin at the back. With a little practice, mounting is quite easy. The best mounting glue is gum tragacanth, which can be bought from a chemist or online. Only small amounts are needed. About 8 g of the gum is placed in a suitable container and water is then added, a little at a time, with continual stirring, until about 15 g has been added. It will keep for a long time if a few drops of preservative, such as isopropanol are stirred in. Alternatively, entomological suppliers sell ready-made gum for mounting beetles as 'Coleoptera mounting agent'.

Every experienced ant worker has their own idiosyncratic way of mounting ants. One method is as follows. The freshly killed ant is placed upside down on the work surface with the ant's head facing away. A fine paintbrush, dipped in gum, is applied, gently, to the ant's underside. This will pick the ant up. Every effort should be made to ensure that the legs are not caught up in the gum. The brush is gently twisted to bring the animal right way up and the ant is lowered onto the prepared card. The brush is then slipped out backwards, leaving the ant on the card. The legs (and wings if present) may need some final arrangement, and this is best done with a fine mounted needle, the fine brush or, in the case of very small specimens, a very fine pin (a minuten) held firmly in curved or entomological forceps or pushed into the rubber on top of a pencil. Alternatively, the ant may be mounted on its side. This allows access to some characters not easily seen in ants mounted ventral side down. An entomological pin can now be pushed firmly through the wide rear margin of the card, so that the card is about 1 cm below the top of the pin. This leaves the rest of the shaft of the pin, below the

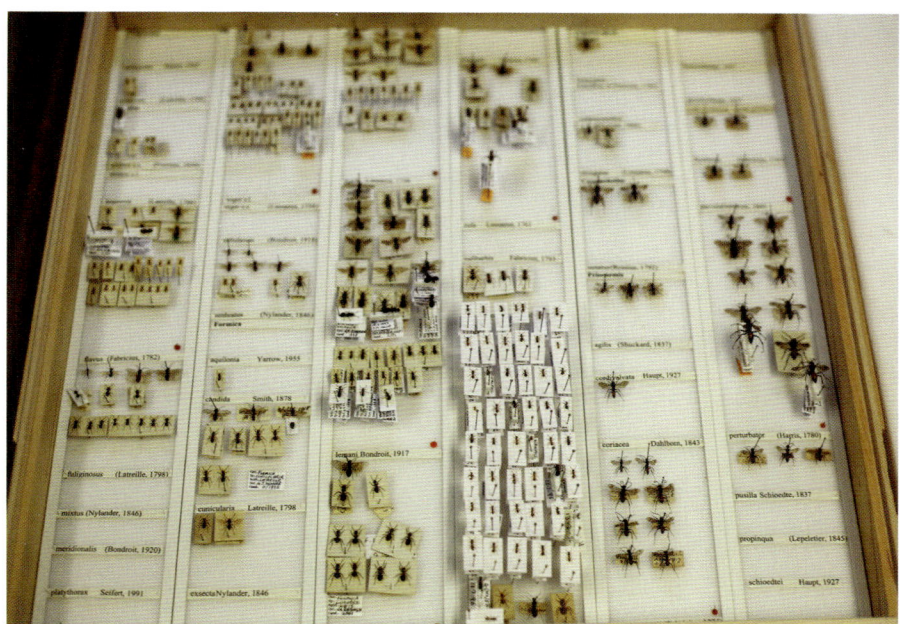

Fig. 7.13 Ants mounted on cards

card, for labels. Suppliers of entomological pins are listed on p. 302. There is a bewildering array of sizes. Do not use English or AA (headless) pins. It is not so important what size pin is used if it is being pushed through a card, as long as it will go through. In the end, it is a matter of preference, but a medium pin is recommended, such as no. 5. It is customary to use two labels. The upper one shows essential details: date of capture, locality, method and name of collector. The lower label shows the Latin name of the ant (Fig. 7.13).

An alternative and increasingly popular way to mount ants is to 'point' them. In this method, the ant is glued to the tip of a triangle of card. This permits examination of the underside of the specimen, which may occasionally be necessary. For this reason, the glue should touch as little as possible (see video at https://youtu.be/OZuTyQd_UJM). The technique takes much less time and is preferred by workers mounting many specimens for research. However, it does not produce such an attractive finished product (Figs 7.14, 7.15).

The whole collection is best kept in a storage box lined with cork or expanded polyethylene foam (Plastazote) (suppliers p. 301). Quite good storage boxes can be made much more cheaply from any close-fitting box, deep enough to accommodate the pins and with a firmly attached floor

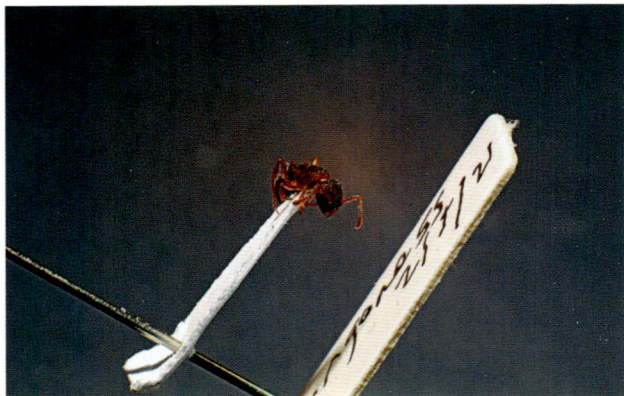

Fig. 7.14 A pointed ant

of cork or Plastazote. The minute amount of soft tissue in the ant's body will dry and shrink, leaving the hard exoskeleton which, with care, will survive for decades or even centuries. The biggest problems are damp (which encourages mould growth) and attack by scavenging insects and mites. Mould can be avoided by storing the collection in a dry place. To prevent insect damage – such as that caused by carpet beetles – the boxes should be airtight and ideally kept inside ziplocked bags. The preservation of insects in storage

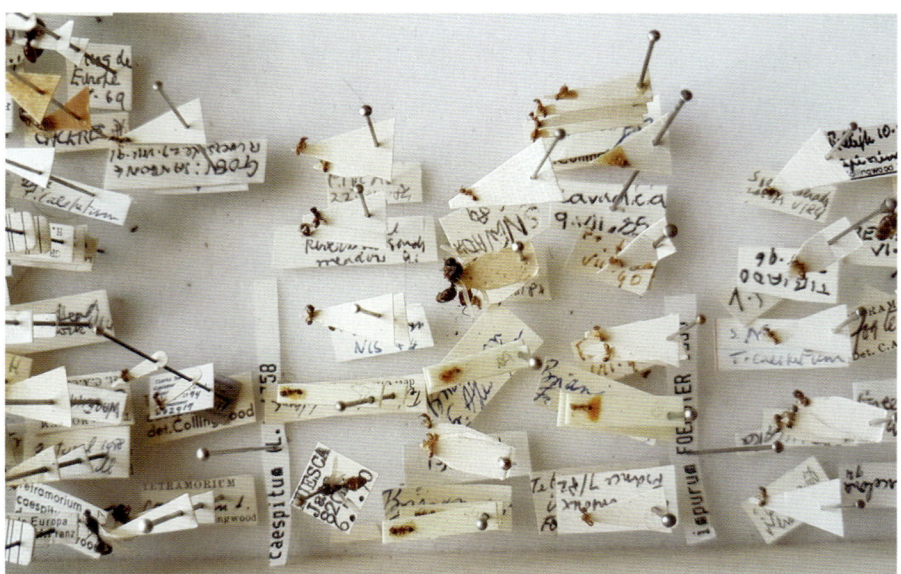

Fig. 7.15 A collection of pointed ants

boxes using chemicals is now generally avoided due to the toxic nature of these chemicals. Most museums these days operate a regime of regular inspection and freezing to keep such problems to a minimum. Mothball packs (e.g. Zensect, available online) can be secured inside if desired.

Larger numbers of specimens are best stored in 70% isopropanol in small tubes. This may be necessary, for example, if one is collecting a series to show variation in worker ants from the same nest. Hölldobler & Wilson (1990) recommend that 20 specimens of each caste are collected, if possible, together with about 20 larvae, and are stored in small vials. These are easily carried in the field and specimens can be put in them immediately. As with dried specimens, those preserved in fluid must be properly labelled with all relevant data. An unlabelled specimen is valueless. The labels should be written in pencil on stiff paper or card and put inside the vial. Isopropanol tends to evaporate eventually, even in a well-sealed container. To prevent this, the vials may be stored submerged in 70% isopropanol in a large wide-mouthed jar with a well-fitting lid.

7.3 Culturing ants

Rearing ants indoors is generally easy, useful and interesting. By keeping colonies in formicaria it is possible to extend the period during which research can be carried out in temperate latitudes, where ants out of doors hibernate in winter. Ant culture has a long history. An early formicarium was described by White (1895). Many modifications and refinements have been made since then, but the basic principles remain the same. The ants must be enclosed to prevent escape, the founding colony must possess a functional queen, there must be provision for feeding the ants, the nest must remain moist for as long as possible, and the ants must have access to a dark region of the formicarium.

There are effectively two types of nests, at least for the smaller species. Lubbock described a glass observation nest in which the ants are confined to a very thin layer of soil between two horizontal glass plates, held apart with slips of wood, one of which is loose and slightly short to leave a door (Fig. 7.16). In contrast, 'Janet nests' contain no soil (Fig. 7.17). Instead of being constructed by the ants in soil, the chambers and galleries are pre-formed in some suitable material. In its original form, this nest consists of four horizontal chambers of plaster of Paris. Three are interconnected by galleries. The fourth acts as a water trough, keeping the others moist by capillary action through the porous plaster (Skaife 1961).

Fig. 7.16 A modern Lubbock-type nest

Fig. 7.17 A Janet-type nest, the Antcube (Antstore, Germany)

Many modifications of these two basic types have been described and used over the last century or so (for instance Wheeler 1910; Donisthorpe 1927a; Skaife 1961; Hölldobler & Wilson 1990) and there is still room for innovation in this area. Modern materials and artefacts provide opportunities for the development of new types of nests, although generally based on the basic Lubbock or Janet design.

Perspex, Ytong, cork and acrylic materials, amongst others, have all been used in modern nests. A recent development is in the field of 3D printing. Some companies are offering 3D-printed nests in very comprehensive modular systems (Fig. 7.18). It is also possible to design and print your own.

The types of nests described here are satisfactory for most British species, but not for very big colonies of large species, such as the wood ants. These can be housed in a large wooden or metal box. The biggest problem is keeping them in. Escape can be prevented by standing the whole box in a trough of water or smearing a layer of fruit-tree banding grease around the rim, but many of the workers will drown or become stuck and die. Furthermore, the grease may become bridged by dead bodies, across which other workers may escape. Renewal of the grease at intervals will solve this problem. Again, there is scope for designing new ways of keeping these larger species (Czechowski & Pisarski 1992).

The food provided for the ants will vary according to species, although Ettershank (1967) invented a diet that would suit most ants. Some of his components are difficult to obtain today, so a version using everyday materials is quoted here. This was invented by Bhatkar & Whitcomb

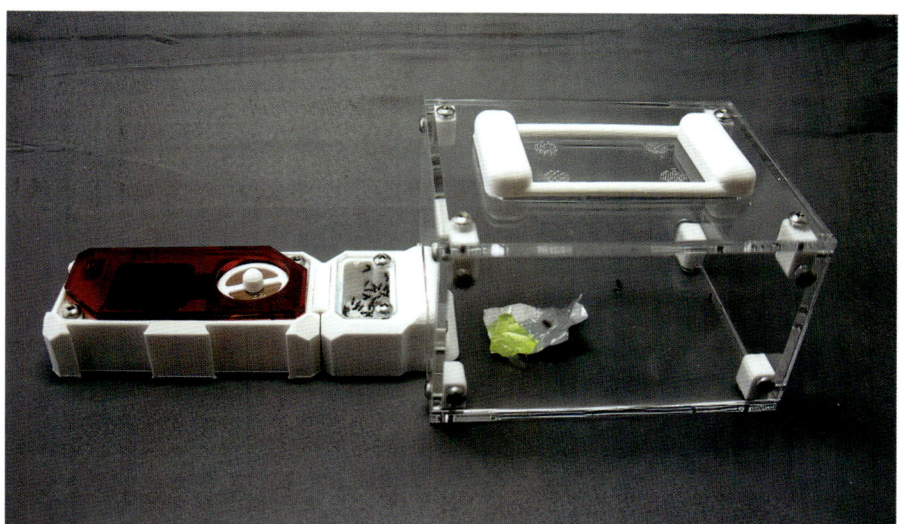

Fig. 7.18 A WaKooshi nest with outworld

(1970) and is used regularly by, for example, Hölldobler & Wilson (1990).

Ingredients for ant diet
1 egg
1 g vitamins*
98 g honey
5 g agar powder (often sold as vegan gelatine at supermarkets or online)
1 g minerals and salts*
500 ml water

Dissolve the agar by adding it gradually to 250 ml boiling water, stirring constantly. Allow to cool but not set at this stage. Mix 250 ml water, honey, vitamins, minerals and egg until smooth, using a liquidiser. Add the cooled agar to this with constant stirring. Pour this final mixture into shallow dishes such as Petri dishes. Store in a refrigerator, where it will form a jelly that can be cut into small blocks for use. This diet should be supplemented with such things as fragments of freshly killed mealworms, cockroaches or crickets, or whole fruit flies. More simply, colonies will often survive

---

* In the original paper of Bhatkar & Whitcomb (1970), McKesson Bexel capsules were used. These are no longer available but there are many substitutes online, such as Multivitamins & Minerals Capsules from Just Vitamins.

quite well on sugar water or honey water supplemented with insect fragments. The water is best supplied in a tube with a cotton wool wick, at which the ants come and drink. Keller *et al.* (1989) suggest another alternative.

Recently, much effort has been devoted to the study of diets for ants of economic importance, such as fire ants, *Solenopsis invicta* (Gavilanez-Slone & Porter 2014), and greenhead ants, *Rhytidoponera metallica* (Dussutour & Simpson 2008). Ideas in these papers could be tried and adapted for species in Britain.

Good starting references for more details on all aspects of ant culture are Hayes (2018) and Luckhurst (2017).

Once a colony of ants is established in an observation nest, there is much scope for detailed study. Behaviour during chamber-construction and gallery-building can be observed in a Lubbock-type nest. Janet and Lubbock nests are both suitable for observations on brood and queen care, and food exchange behaviour within the nest. In observation nests with foraging arenas (generally referred to as 'outworlds'), foraging behaviour, food finding and recruitment can be investigated. Potted plants infested with aphids can be placed in larger arenas, providing an opportunity to gather information on tending behaviour and honeydew intake. Throughout all such work it must be borne in mind that this is an artificial situation for the ants. The study of observation nests must always be treated as a supplement to work in the field, not as a substitute for it.

7.4 Estimating colony size

It is often necessary to estimate colony size. For animals as mobile as ants, the mark-recapture or Lincoln index technique would seem to be an appropriate, non-destructive method (Chen & Robinson 2013). Several individuals are captured – alive and undamaged – and marked in some way. A known number of such marked individuals is then released back into the population. At some later time, the population is sampled at random, and the number of captures and recaptures (marked individuals caught) is recorded. The total population size can then be estimated from:

$$N = \frac{M \times n}{m}$$

Where:

N = estimate of population size
M = number of marked individuals released

n = total number caught on second sampling occasion

m = number of marked individuals caught on second sampling occasion

This technique is easy to use, but time consuming because the accuracy of the estimate increases with the number of individuals marked. It involves several assumptions, which must be appreciated before the results are interpreted. The standard error (SE) of the estimate is given by:

$$SE = \sqrt{\frac{M^2 \times n \times (n - m)}{m^3}}$$

(from Wheater & Cook 2002, where a worked example can be found).

The assumptions are:

1. The marks do not affect the animals in any way, and they are not lost during the period of the investigation.
2. The marked animals become mixed within the population.
3. The population is sampled randomly.
4. There is no birth or death.
5. There is no emigration or immigration.

It is the second of these that causes most problems with ants. The whole ant population does not mix freely, because some workers are restricted to the nest. Any mark-recapture exercise will give only an estimate of the numbers of foragers and this number will change with time of day, weather conditions and season. It is possible to estimate forager numbers on trails, where individual ants tend to remain on the same trail (Holt 1955). If the mark-recapture technique is to be used, a suitable marking method is needed. For some of the bigger species, such as the wood ants, the gaster can be marked with a small dot of quick-drying paint, such as cellulose dope, which is available online. It does not seem to affect survival in any obvious way, but the effects of marking should be examined in specific situations, by a combination of laboratory investigations (checking such things as toxicity of marking agent) and field experiments (checking, for instance, if a marked animal is more or less likely to be eaten, or to be able to catch food). Other marking methods include dusting with fluorescent powder (which shows up in ultra-violet light), powdered dyes (revealed by putting marked ants on wet white filter paper and, if necessary, applying a drop of acetone) or a 7% solution of phenolphthalein in acetone (which turns purple when recaptured ants are put in 1% sodium hydroxide solution), or even labelling with radioactive tracers. These have the great advantage of not being visible to any potential predator or prey, or indeed to the experimenter when performing the second capture, but their use requires special precautions. Marking methods and the Lincoln index technique are reviewed by Chalmers & Parker (1989), Henderson (2021) and Wheater & Cook (2002).

7.5 Territory mapping

Probably all ant species are territorial. Even if they do not always show vigorous defence of an established area (an absolute territory) (Hölldobler & Wilson 1990), they will defend some parts of the foraging area (spatiotemporal territory). In either case, some assessment of territory boundaries is often required in a study of a species or ant community. One method takes advantage of the worker's ability to recognise friend or foe. In wood ants, for example, workers from one colony, removed and then replaced, elicit no more than mild interest from fellow workers. In contrast, if workers from a neighbouring colony are introduced, a vigorous reaction will be observed. A worker from a different colony will be attacked by an increasing number of the original colony workers and is eventually killed and carried off to the home nest. Similar experiments have been carried out on many other species (Hölldobler 1979). With care, this response can be used to map territorial boundaries quite accurately. Ants can initially be transferred from known points on the ground or within vegetation to establish which colony controls that spot. A further sophistication is to set out baits of 2% sugar solution in selected and known places. Aggression tests can then be used to see which, of all possible colonies, controls the place where the bait was placed. An accurate map of the study site, or grid made of string or other material, is needed for this kind of work. Techniques for mapping are described by Gilbertson *et al.* (1985). In a study of *Formica rufa*, a simple method was used for mapping trees in a woodland area. A tree was chosen at random on the edge of the wood and numbered 1. Compass bearings were taken with a sighting compass, and distances were measured from this tree to a few others. The data were recorded, and the trees numbered as they were done. When the walking between trees got tiresome, one of the mapped trees was taken as a new base point and mapping continued from this tree. In this way, the observer can leap-frog through the wood and many trees can be mapped accurately and very quickly. The data can be transferred to a large sheet of graph paper with protractor and ruler. This method could be adapted to map other objects on a much smaller scale. A map produced with *Formica rufa* territory boundaries and trails is shown in Fig. 2.13. If the resources are available, mapping can be carried out with a drone and graphics software (Calvo 2018).

7.6 Feeding habits

It is relatively easy to assess the food income of a colony in trail-forming ants such as *Formica* species of the wood ant group (*F. rufa, F. lugubris, F. aquilonia, F. pratensis*) or *Lasius fuliginosus*. Food coming in along trails can be observed directly. A problem is, however, that if food is confiscated from a returning worker, then other workers become agitated in response to alarm pheromones (p. 77). Less disturbing methods have been devised. In one example (Skinner 1980a,b), wood ant nests were enclosed in an ant-proof fence and the movements of foragers in and out were restricted to specific entry and exit points. The ant-proofing was achieved by burying the fence panels in a trench filled with old engine oil and by smearing its top with tree-banding grease. The entrance and exits were at the tops of ramps which ended in midair so that incoming and outgoing flows could be split. All incoming ants fell into a box from which the only exit was through holes just big enough to allow ants out. Anything they were carrying was left in the box and could be examined at leisure (Figs 7.19–7.21).

Another way to get information about food intake is to weigh departing and returning workers. In a study of wood ants, for example, it was shown that the mean mass of workers leaving a nest on a particular trail was significantly

Fig. 7.19 A metal fence around a wood-ant nest, with fruit-tree banding grease at the top

Fig. 7.20 Close view of ingoing and outgoing ramps

Fig. 7.21 Ramp system established around wood-ant nest, with collection box inside the fence

Fig. 7.22 Two workers of *Formica aquilonia* returning to the nest with abdomens swollen (replete) with honeydew

less than that of those returning with no food item in their mandibles, but with a swollen gaster. The difference was due to aphid honeydew in the apparently unladen returning workers (Fig. 7.22). Simple weighing can provide valuable information.

7.7 Ant activity patterns

For many years, ecologists and behavioural biologists have been interested in animal activity patterns and have sought to interpret them in terms of 24-hour (diel) rhythms or variations related to weather, food supply and other factors. In the ants, many opportunities exist for such work. In the field, activity can be investigated, simply but laboriously, by camping near a nest and making regular assessments of movements. For instance, the numbers of workers leaving or entering the nest per unit time can be counted on a trail, or the number of ants visible can be counted in a known area or during a measured searching period. Some of this can be automated. Automatic ant counters have been designed in the past (Skinner 1980a,b; Dibley & Lewis 1972; Noda *et al.* 2006). The availability of modern electronic components, computer logging facilities and cheap video recording gives scope for developing new techniques. Videos of trails have been analysed using machine learning or pixel-change trackers to extract information afterwards. The analysis of videos, which can be filmed on any basic camera or a mobile phone, is a promising area for investigation. Sanint reports on an ant counter (https://sites.google.com/site/antcounter/?pli=1) which can be used in this way.

7.8 Effects of ants on other animals

Tended and predated herbivores

As we have seen (p. 71), ants tend aphids and they kill and eat many other herbivorous animals, including non-tended aphids. The impact of the ants can be investigated simply by excluding ants from vegetation on which they would otherwise forage and comparing the success of selected herbivores in this situation with their success in nearby unprotected plots. Sticky tree-banding greases, available commercially for excluding wingless moths from fruit trees, have been useful for excluding ants from trees and shrubs. Research of this kind can also be done using laboratory colonies in formicaria with large foraging arenas.

Ground-dwelling species

Ants may affect other animals living on the ground either by direct predation or by competition. Pitfall trapping (p. 279), although a technique with some problems when used for estimating numbers of ground-living animals, can be employed as a means of investigating these effects. If traps are set out around the nest, ground-living animals are caught. If their numbers are greater outside than inside a nest territory or increase with distance from the nest, predation or competition may be implicated. To find out if either of these is operating may require further experiments or observations. For example, large counts of a particular species as incoming prey may link in with depletion of the numbers of that species near the nest. On the other hand, if a species was never brought in as prey but was much reduced in numbers near the nest, competition would be indicated. Further studies might involve supplying extra food in the ant territories to see whether the reduction of competition for food would allow the numbers of other species to increase.

Myrmecophiles

The subject of species living in association with ants would warrant another book. However, the area is ripe for investigation. Observations can be made by simply watching nests and looking for non-ant species. For example, Robinson (1998) was able to record the tiny spider *Thyreosthenius biovatus* with *F. lugubris* by the simple expedient of looking. In another example, one of the authors (GJS) and E. Robinson observed larvae of the ant-associated beetle *Clytra quadripunctata* on nests of *F. lugubris*. This species is fascinating. The female coats its eggs in mucilage and faeces to form a case and drops

Fig. 7.23 *Clytra quadripunctata* larva with replete *F. lugubris* worker

reflex bleeding
A defensive behaviour
in some insects where a
noxious fluid is produced
if it is threatened.

them near nests of wood ants. The ants bring them to the nest where they hatch and feed on plant debris, protecting themselves by adding more faeces to the case they start with (Fig. 7.23). When the adult emerges it leaves the nest, exuding fluid by reflex bleeding if approached by an ant. The larvae are protected from the cold weather of the winter and gain an advantage from this.

As indicated in Chapter 3, much could be learned by observation of nests that harbour these ant-associated species. For example, the list of just six true guests on p. 62 (itself based on a list that Donisthorpe compiled nearly 100 years ago) includes only two species where the number of records on the National Biodiversity Network exceeds double figures. One feels they must occur elsewhere and that we just need to look for them. This is a job for anyone keen and persistent.

7.9 Presenting your findings

It is relatively easy to make your discoveries available to the wider world. In the first instance, any ant records that you feel may be of interest should be communicated to the Bees, Wasps and Ants Recording Society (BWARS at www.bwars.com). Very comprehensive instructions of how to submit records are available on the site. Records can also be submitted to iRecord at www.brc.ac.uk/irecord. This can also be used for records of myrmecophiles. It is also a good idea to send records to the relevant local environmental records centre (https://www.alerc.org.uk). Your findings could be disseminated via a blog or a local natural history society newsletter; many options exist online. There are also numerous links on Facebook to groups interested in most animals.

For more substantial findings, you may wish to try and publish in a journal. The BWARS newsletter would be one option. Other possibilities include *The Entomologist's Monthly Magazine, Myrmecological News, British Journal of Entomology and Natural History* and *The Entomologist's Record and Journal of Variation*. If your finding is concerned with a myrmecophile, the relevant journal for the group might be the best place. For example, if you found a beetle of interest then you might consider *The Coleopterist*.

When writing up any piece of work, you must do so honestly and with full disclosure of your findings. Wheater & Cook (2003) and Chalmers & Parker (1989) are both useful guides to writing up and for help with any statistical issues where your observations have a numerical aspect.

8 Useful addresses, links and other resources

8.1 Equipment suppliers

The following is a selection; a web search will find other suppliers too, especially of ant-keeping equipment.

Ant Keeping Depot. 18–22 Bridge Street, Eltham, Victoria 3095, Australia (warehouse only).
Email: info@antkeepingdepot.com
www.antkeepingdepot.com
Suppliers of formicaria, sundries and live ants. They sell products from other companies, so a very wide range. Australia-based but they ship internationally at reasonable rates.

Ant Antics. 1 Priory St., Carmarthen SA31 1LS.
Email: hello@antantics.wales
www.antantics.wales
Suppliers of formicaria, a wide range of sundries and live ants.

Brunel Microscopes Ltd. Unit 2, Vincients Road, Bumpers Farm Industrial Estate, Chippenham, Wiltshire, SN14 6NQ.
Tel: 01249 462655
Email: mail@brunelmicroscopes.co.uk
www.brunelmicroscopes.co.uk
Wide range of both new and used microscopes and hand lenses.

D.J. & D. Henshaw. 34 Rounton Rd, Waltham Abbey, Essex, EN9 3AR.
Tel: 01992 717663
Email: djhagro@aol.com
Entomological and microscopy supplies, magnifiers, specimen tubes, pooters, entomological pins, Plastazote (for lining storeboxes), data label printing service, silicone tubing for pooters.

NHBS Ltd. 1–6 The Stables, Ford Road, Totnes, Devon, TQ9 5LE.
Tel: 01803 865913
Email: customer.services@nhbs.com
www.nhbs.com
Supplier of books and survey equipment.

Philip Harris. Philip Harris Education, 2 Gregory Street, Hyde, Cheshire, SK14 4TH.
Tel: 03451 204520
Email: enquiries@philipharris.co.uk
www.philipharris.co.uk
Field and laboratory equipment supplier.

RS Components. Birchington Road, Corby, Northamptonshire, NN17 9RS (with 16 local outlets around the UK).
Email: online form at https://uk.rs-online.com/web/
Wide range of electronics components and a source of PTFE (Rocol Lubricant PTFE).

WaKooshi.
Tel: 01628 412510
Email: sales@wakooshi.com
www.wakooshi.com
Suppliers of modular formicaria systems, sundries such as ant food (both artificial and live), heating, escape proofing, live ants etc. and publishes *Ant Keeper Magazine*.

Watkins and Doncaster. Golderfield, Pudleston, Leominster, Herefordshire, HR6 0RG. Postal address: PO Box 114, Leominster, HR6 6BS.
Tel: 0333 8003133
Email: online contact via the form at www.watdon.co.uk
www.watdon.co.uk
Suppliers of a very wide range of entomological equipment. Ethyl acetate can be purchased from them.

8.2 Book suppliers

The following is a selection; a web search will find other suppliers.

AbeBooks. AbeBooks Europe GmbH, Marcel-Breuer-Str. 12, 80807 München, Germany.
Email: info@abebooks.co.uk
www.abebooks.co.uk
Web-based used-book site. Vast coverage means you can find most out-of-print books here.

NHBS Ltd. 1–6 The Stables, Ford Road, Totnes, Devon, TQ9 5LE.
Tel: 01803 865913
Email: customer.services@nhbs.com
www.nhbs.com
Supplier of books and survey equipment.

Pemberley Books. 18 Bathurst Walk, Iver, Buckinghamshire, SL0 9AZ.

Tel: 01753 631114

Email: orders@pemberleybooks.com or online contact via the form at www.pemberleybooks.com

www.pemberleybooks.com

Suppliers of a wide range of both new and used natural history books.

8.3 Societies and interest groups

Amateur Entomologists' Society (AES). PO Box 8774, London, SW7 5ZG.

Email: contact@amentsoc.org

www.amentsoc.org

The Amateur Entomologists' Society is one of the UK's leading organisations for people interested in insects. The society produces four publications for members together with a series of books.

Bees, Wasps and Ants Recording Society (BWARS).

Email: online contact via the form at www.bwars.com

www.bwars.com

A society whose main aim is to collate records of aculeates (ants, bees and wasps). The website has maps, species accounts, identification guides and a forum where you can ask questions. They also run courses on various aspects of aculeate biology, mainly identification.

Biological Records Centre. UK Centre for Ecology & Hydrology, Wallingford, Oxfordshire, OX10 8BB

Tel: 01491 692357

Email: brc@ceh.ac.uk

A national focus in the UK for terrestrial and freshwater species recording, established in 1964. They work closely with the voluntary recording community, principally through support of national recording schemes and societies.

British Entomological and Natural History Society (BENHS). The Pelham-Clinton Building, Dinton Pastures Country Park, Davis Street, Hurst, Reading, Berkshire, RG10 0TH.

Email: enquiries@benhs.org.uk

www.benhs.org.uk

Buglife (Invertebrate Conservation Trust). Bug House, Ham
Lane, Orton Waterville, Peterborough, Cambridgeshire,
PE2 5UU.
Email: info@buglife.org.uk
www.buglife.org.uk
Buglife is involved in the conservation of all invertebrates
via a range of campaigns.

International Union for the Study of Social Insects (IUSSI).
Email (for membership questions): secretary.nweu.iussi@
gmail.com
www.iussi.org
Promotes the study of social insects worldwide. Publishes
the journal *Insectes Sociaux*.

The National Biodiversity Network. Unit F, 14 – 18 St. Mary's
Gate, Lace Market, Nottingham, NG1 1PF.
Tel: 0115 850 0177
Email: support@nbn.org.uk
www.nbn.org.uk.
A portal for the sharing and inspection of records. The
site features distribution maps of thousands of species.

Royal Entomological Society. The Mansion House, Chiswell
Green Lane, St Albans, Hertfordshire, AL2 3NS, UK.
Tel: 01727 899387
Email: info@royensoc.co.uk
www.royensoc.co.uk
The society covers all aspects of entomology and has a
superb library.

9 References and further reading

Since there are relatively few ant species in the British Isles, it has been possible to treat them comprehensively in this book. There has not been another book on the ants of Britain and Ireland since the publication of the first edition of this book in 1996 until the volume by Jones (2022), which keys out only workers. Seifert (2018) and Lebas *et al.* (2019) treat the British and Irish ants, in English, within the wider context of the rest of Europe. Czechowski *et al.* (2012) is also useful as it is, again available in English, and covers British and Irish species. Seifert (2018) is comprehensive, with an extensive reference list.

Other works are now out of date but might still be useful for their illustrations and descriptions. These include Bolton & Collingwood (1975), for long the definitive key, Collingwood (1979) for Fennoscandia but covering most British and Irish species, and Brian (1976). Donisthorpe (first edition 1915, second edition 1927a) is also a source of detailed information not readily available elsewhere.

An important resource is the county or regional guide. In the case of ants, they are usually found, along with the wasps and bees, in a volume covering all the aculeate Hymenoptera. To date, those available are:

aculeate Hymenoptera
Ants, bees and wasps that have a sting or had a sting primitively (some ants no longer have a sting). This includes all ants, all bees and the wasps apart from the parasitic wasps.

- Cumbria (Robinson 2005)
- Essex (Harvey 1998)
- Hertfordshire (Attewell 2008)
- Highlands of Scotland, covering the Highland Council local authority area (Macdonald 2013)
- Ireland (Niechoj 2011)
- Kent (Allen 2020)
- Lancashire (Hargreaves & White 2021)
- Leicestershire and Rutland (Gamble 2020)
- Norfolk (Wells 2018)
- Shropshire (Jones & Cheeseborough 2014)
- Surrey (Pontin 2005)
- Yorkshire (Archer 2002)

A search online will reveal much detailed information for some counties. For example, Essex at https://www.essexfieldclub.org.uk/portal.php and Nottinghamshire at http://www.eakringbirds.com/eakringbirds2/insectsants.htm), but rather generic material for others.

Whilst all the above works include more than just identification keys and pictures, there are further specialist books on the biology and ecology of ants. These include Stockan & Robinson (2016) (covering all European and North American members of the genus *Formica* regarded as wood ants), Lach *et al.* (2010), and Hölldobler & Wilson (1990, 2009).

Beyond this, information can be sought in journals, some of which are dedicated to all insects, some the social Hymenoptera and some even just ants. These include:

- *The Entomologist's Monthly Magazine* and *Entomologist's Gazette* (both published by Pemberley Books, see p. 303, covers all insect orders).
- *The Entomologist's Record and Journal of Variation* (published by the Amateur Entomologists' Society, covers all insect orders).
- *Myrmecological News* (a free online peer-reviewed journal devoted to ants).
- *BWARS Newsletter* (free to BWARS members, see p. 303, covers ants, bees and wasps).
- *Insectes Sociaux* (the journal of the International Union for the Study of Social Insects or IUSSI, covers all social insects).

Bibliography

Allen, G.W. (2020) *Bees, Wasps and Ants of Kent*. 2nd edition. Orpington: Kent Field Club.

Archer, M.E. (2002) *The Wasps, Ants and Bees of Watsonian Yorkshire*. Weymouth: Yorkshire Naturalists' Union.

Andersen, A.N. & Majer, J.D. (2004) Ants show the way down under: invertebrates as bioindicators in land management. *Frontiers in Ecology and Environment* 2: 291–298. https://doi.org/10.1890/1540-9295(2004)002[0292:ASTWDU]2.0.CO;2

Attewell, P. (2019) Wall ants and wood ants. *Hertfordshire Naturalist* 51: 30–33.

Attewell, P. (2020) Wood ants and other ants, 2020/2021. *Hertfordshire Naturalist* 53: 35–40.

Attewell, P.J. (2008) Ants, 2007, with county checklist. *Hertfordshire Naturalist* 40: 16–22. https://www.hnhs.org/archivedisplay?page=H-40-1-18

Attewell, P.J. & Wagner, H.C. (2019) *Tetramorium impurum* Foerster (Hymenoptera: Formicidae), first record for Guernsey and the Channel Islands. *British Journal of Entomology and Natural History* 32: 287–296.

Baxter, F.P. & Hole, F.D. (1966) Ant (*Formica cinerea*) pedoturbation in a prairie soil. *Soil Science Society of America Proceedings* 31: 425–428. https://doi.org/10.2136/sssaj1967.03615995003100030036x

Beattie, A.J. (1985) *The Evolutionary Ecology of Ant–Plant Mutualisms*. Cambridge: Cambridge University Press. https://doi.org/10.1017/CBO9780511721878

Beugnon, G. & Fourcassie, V.J. (1988) How do red wood ants orient during diurnal and nocturnal foraging in a three dimensional system? II. Field experiments. *Insectes Sociaux* 35: 106–124. https://doi.org/10.1007/BF02224142

Benyon, R., Stevenson, S., Griffiths, J. & Atwood, A. (2012) UK Post-2010 Biodiversity Framework. Found at https://hub.jncc.gov.uk/assets/587024ff-864f-4d1d-a669-f38cb448abdc#UK-Post2010-Biodiversity-Framework-2012.pdf

Bhatkar, A.P. & Whitcomb, W.H. (1970) Artificial diet for rearing various species of ants. *Florida Entomologist* 53: 229–232. https://doi.org/10.2307/3493193

Bier, K. (1954) Über den Einfluss der Königin auf die Arbeiterinnenfertilität im Ameisenstaat. *Insectes Sociaux* 1: 7–19. https://doi.org/10.1007/BF02223147

Billen, J.P.J. (1984) Stratification in the nest of the slave-making ant *Formica sanguinea* Latreille, 1798 (Hymenoptera, Formicidae). *Annales de la Société royale zoologique de Belgique* 114: 215–225.

Boer, P. (2008) Observations of summit disease in *Formica rufa* LINNAEUS, 1761 (Hymenoptera: Formicidae). *Myrmecological News* 11: 63–66.

Bolton, B. & Collingwood, C. (1975) *Hymenoptera: Formicidae*. Handbooks for the Identification of British Insects Vol. 6 Pt 3(c). London: Royal Entomological Society.

Boomsma, J.J. (2009) Lifetime monogamy and the evolution of eusociality. *Philosophical Transactions of the Royal Society B Biological Sciences* 364: 3191–3207. https://doi.org/10.1098/rstb.2009.0101

Bos, N., Kankaanpää-Kukkonen, V., Freitak, D., Stucki, D. & Sundström, L. (2019) Comparison of twelve ant species and their susceptibility to fungal infection. *Insects* 10: 271. https://doi.org/10.3390/insects10090271

Brangham, A.N. (1938) Additions to the wild fauna and flora of the Royal Botanic Gardens, Kew: XVIII. The Ants of the Royal Botanic Gardens, Kew. *Bulletin of Miscellaneous Information (Royal Botanic Gardens, Kew)* 1938: 390–396. https://doi.org/10.2307/4111508

Breen, J. (2014) Species dossier, range and distribution data for the Hairy Wood Ant, *Formica lugubris*, in Ireland. Irish Wildlife Manuals, No. 68. National Parks and Wildlife Service, Department of the Arts, Heritage and the Gaeltacht. https://www.npws.ie/sites/default/files/publications/pdf/IWM68.pdf

Brian, M.V. & Brian, A.D. (1955) On the two forms macrogyna and microgyna of the ant *Myrmica rubra* L. *Evolution* 9: 280–290. https://doi.org/10.1111/j.1558-5646.1955.tb01537.x

Brian, M.V. & Carr, C.A.H. (1960) The influence of the queen on brood rearing in ants of the genus *Myrmica*. *Journal of Insect Physiology* 5: 81–94. https://doi.org/10.1016/0022-1910(60)90034-2

Brian, M.V. (1955) Food collection by a Scottish ant community. *Journal of Animal Ecology* 24: 336–351. https://doi.org/10.2307/1717

Brian, M.V. (1964) Ant distribution in a southern English heath. *Journal of Animal Ecology* 33: 451–461. https://doi.org/10.2307/2565

Brian, M.V. (1965) *Social Insect Populations*. London: Academic Press.

Brian, M.V. (1977) *Ants*. New Naturalist 59. London: Collins.

Brian, M.V. (1983) *Social Insects: Ecology & Behavioural Biology*. London: Chapman & Hall. https://doi.org/10.1007/978-94-009-5915-6

Bristow, C.M. (1984) Differential benefits from ant attendance to two species of Homoptera on New York ironweed. *Journal of Animal Ecology* 53: 715–726. https://doi.org/10.2307/4654

Brooks, J.L. (1942) Notes on the ecology and the occurrence in America of the mymecophilous sowbug *Platyarthrus hoffmannseggi*. *Ecology* 23: 427–437. https://doi.org/10.2307/1930129

Buschinger, A. (1968) Mono- und Polygynie bei Arten der Gattung *Leptothorax* Mayr (Hymenoptera, Formicidae). *Insectes Sociaux* 15: 217–226. https://doi.org/10.1007/BF02225844

Çamlitepe, Y. & Stradling, D.J. (1995) Wood ants orient to magnetic fields. *Proceedings of the Royal Society B: Biological Sciences* 261: 37–41. https://doi.org/10.1098/rspb.1995.0114

Calvo, K. (2018) *So You Want to Create Maps Using Drones?* Blurb (www.blurb.com).

Carr, C.A.H. (1962) Further studies on the influence of the queen in ants of the genus *Myrmica*. *Insectes Sociaux* 9: 197–211. https://doi.org/10.1007/BF02329893

Carroll, S. (2008) Formica exsecta *in Devon in 2008*. Midhurst: Hymettus.

Carthy, J.D. (1951) The orientation of two allied species of British ant, ii. odour trail laying and following in *Acanthomyops* (*Lasius*) *fuliginosus*. *Behaviour* 3: 304–318. https://doi.org/10.1163/156853951X00313

Chalmers, N. & Parker, P. (1989) *The OU Project Guide: Fieldwork and Statistics for Ecological Projects*. London: Open University/Field Studies Council.

Chen, Y.H. & Robinson, E.J.H. (2013) A comparison of mark–release–recapture methods for estimating colony size in the wood ant *Formica lugubris*. *Insectes Sociaux* 60: 351–359. https://doi.org/10.1007/s00040-013-0300-z

Chomicki, G. & Renner, S.S. (2017) The interactions of ants with their biotic environment. *Proceedings of the Royal Entomological Society of London B* 284: 20170013. https://doi.org/10.1098/rspb.2017.0013

Collingwood, C.A. & Hughes, J. (1987) Ant species in Yorkshire. *Naturalist* (Leeds) 112: 95–101.

Collingwood, C.A. (1979). *The Formicidae (Hymenoptera) of Fennoscandia and Denmark*. Fauna Entomologica Scandinavica Volume 8. Leiden: Brill. https://doi.org/10.1163/9789004273337

Czechowski, W., Radchenko, A.G., Czechowska, W. & Vepsalainen, K. (2012) *The Ants of Poland with Reference to the Myrmecofauna of Europe*. Warszawa: Museum and Institute of Zoology of the Polish Academy of Sciences and Natura optima dux Foundation.

Czechowski, W. & Pisarski, B. (1992) Laboratory methods for rearing ants. *Memorabilia Zoologica* 45: 1–32.

Czerwinski, Z., Jakubczyk, H. & Petal, J. (1971) Influence of ant hills on meadow soils. *Pedobiologia* 7: 277–285. https://doi.org/10.1016/S0031-4056(23)00472-9

D'Ettore, P. & Lenoir, A. (2010) Nestmate recognition. In Lach, L., Parr, C.L. & Abbott, L.L. (eds). *Ant Ecology* (pp. 194–209). Oxford: Oxford Academic. https://doi.org/10.1093/acprof:oso/9780199544639.003.0011

Darwin, C. (1859) *On the Origin of Species by Means of Natural Selection*. London: Murray.

Davies, N.B., Krebs, J.R. & West, S.A. (2012) *An Introduction to Behavioural Ecology*. 4th edition. Chichester: John Wiley and Sons.

Dibley, G.C. & Lewis, T. (1972) An ant counter and its use in the field. *Entomologia Experimentalis et Applicata* 15: 499–508. https://doi.org/10.1111/j.1570-7458.1972.tb00237.x

Disney, R.H.L. (1983) *Scuttle Flies, Diptera, Phoridae (except* Megaselia*)*. Handbooks for the Identification of British Insects Vol. 10 Pt 6. London: Royal Entomological Society.

Disney, R.H.L. (1994) *Scuttle Flies: The Phoridae*. London: Chapman & Hall. https://doi.org/10.1007/978-94-011-1288-8

Dixon, A. & Thieme, T. (2020) *Aphids on Deciduous Trees*. Naturalists' Handbooks 29. Exeter: Pelagic Publishing.

Dobrzanska, J. (1959) Studies on the division of labour in ants, genus *Formica*. *Acta Biologiae Experimentalis* 19: 57–81.

Dodd, S. (2015) Formica rufibarbis: *Survey and Monitoring Report for 2014*. Woking: Surrey Wildlife Trust.

Domisch, T., Risch, A.C. & Robinson, E.J.H. (2016) Wood ant foraging and mutualism with aphids. In: Stockan, J. & Robinson, E. (eds). *Wood Ant Ecology and Conservation (Ecology, Biodiversity and Conservation)*. Cambridge: Cambridge University Press. https://doi.org/10.1017/CBO9781107261402.008

Donisthorpe, H.St.J.K. (1915) *British Ants*. 1st edition. Plymouth: William Brendon and Son.

Donisthorpe, H.St.J.K. (1927a) *British Ants*. 2nd edition. London: George Routledge and Sons.

Donisthorpe, H.St.J.K. (1927b) *The Guests of British Ants*. London: George Routledge and Sons.

Douglas, J.M. & Sudd, J.H. (1980) Behavioural coordination between an aphid (*Symydobius oblongus* von Heyden; Hemiptera: Callaphidae) and the ant that attends it (*Formica lugubris* Zetterstedt; Hymenoptera: Formicidae): an ethological analysis. *Animal Behaviour* 28: 1127–1139. https://doi.org/10.1016/S0003-3472(80)80102-3

Duff, A.G. (2024) *Beetles of Britain and Ireland. Vol. 2: Staphylinidae*. West Runton: A.G. Duff.

Dumpert, K. (1978) *The Social Biology of Ants* (translated by C. Johnson). London: Pitman.

Dussutour, A. & Simpson, S.J. (2008) Description of a simple synthetic diet for studying nutritional responses in ants. *Insectes Sociaux* 55: 329–333. https://doi.org/10.1007/s00040-008-1008-3

Ehrhardt, S. (1931) Über Arbeitsteilung bei *Myrmica*- und *Messor*-Arten. *Zeitschrift für Morphologie und Ökologie der Tiere* 20: 755–812. https://doi.org/10.1007/BF00407687

El-Ziady, S. & Kennedy, J.S. (1956) Beneficial effects of the common garden ant, *Lasius niger* L., on the black bean aphid. *Aphis fabae* Scopoli. *Proceedings of the Royal Entomological Society of London A* 31: 61–65. https://doi.org/10.1111/j.1365-3032.1956.tb00208.x

Elgert, B. & Rosengren, R. (1977) The guest ant *Formicoxenus nitidulus* follows the scent trail of its wood ant host (Hymenoptera, Formicidae). *Memoranda Societatis pro Fauna et Flora Fennica* 53: 35–38.

Elmes, G. (1974) The spatial distribution of a population of two ant species living in limestone grassland. *Pedobiologia* 14: 412–418. https://doi.org/10.1016/S0031-4056(23)00134-8

Elmes, G. & Wardlaw, J. (1982) A population study of the ants *Myrmica sabuleti* and *Myrmica scabrinodis* living at two sites in the south of England. II. Effect of above-nest vegetation. *Journal of Animal Ecology* 51: 665–680. https://doi.org/10.2307/3990

Errard, C. (1994) Developments of interspecific recognition behavior in the ants *Manica rubida* and *Formica selysi* (Hymenoptera: Formicidae) reared in mixed-species groups. *Journal of Insect Behavior* 7: 83–99. https://doi.org/10.1007/BF01989829

Ettershank, G. (1967) A completely defined synthetic diet for ants (Hym., Formicidae). *Entomologist's Monthly Magazine* 103: 66–67.

Evans, H.C., Elliot, S.L., & Hughes, D.P. (2011) Hidden diversity behind the zombie-ant fungus *Ophiocordyceps unilateralis*: four new species described from carpenter ants in Minas Gerais, Brazil. *PLoS ONE* 6: e17024. https://doi.org/10.1371/journal.pone.0017024

Faber, W. (1967) Beiträge zur Kenntnis sozialparasitischer Ameisen. I. *Lasius* (*Austrolasius* n. sg.) *reginae* n. sp., eine neue temporär sozialparasitische Erdameise aus Österreich (Hym. Formicidae). *Pflanzenschutz Berichte* 36: 73–107.

Falk, S. (1991) *A Review of the Scarce and Threatened Bees, Wasps and Ants of Great Britain*. Research and Survey in Nature Conservation No. 35. Peterborough: Nature Conservancy Council.

Farji-Brener, A.G. & Werenkraut, V. (2017). The effects of ant nests on soil fertility and plant performance: a meta-analysis. *Journal of Animal Ecology* 86: 866–877. https://doi.org/10.1111/1365-2656.12672

Feynman, R.P. (1985) *Surely You're Joking, Mr. Feynman*. New York: Morton.

Fowles, A.P. & Hurford, C. (1996). A monitoring programme for the bog ant *Formica candida* (=*transkaucasica*) on Cors Goch Llanllwch SSSI, Carmarthenshire. Cardiff. *Countryside Council for Wales Natural Science Report* 96/5/4.

Franks, N.R., Healy, K.J. & Byrom, L. (1990) Studies on the relationship between the ant ectoparasite *Antennophorus grandis* (Acarina: Antennophoridae) and its host *Lasius flavus* (Hymenoptera: Formicidae). *Journal of Zoology* 225: 59–70. https://doi.org/10.1111/j.1469-7998.1991.tb03801.x

Frouz, J. & Jilková, V (2008). The effect of ants on soil properties and processes (Hymenoptera: Formicidae). *Myrmecological News* 11: 191–199.

Fry, J.M. (1998) *Foraging Patterns of the Wood Ant* Formica rufa Linnaeus *(Hymenoptera: Formicidae) at Burnham Beeches, Buckinghamshire*. Cranfield University: unpublished PhD thesis. http://dspace.lib.cranfield.ac.uk/handle/1826/7040

Gamble, G. (2020) A Checklist of Ants in VC55 – Leicestershire and Rutland. https://www.naturespot.org.uk/sites/default/files/downloads/A%20Checklist%20of%20Ants%20in%20VC55%20–%20Leicestershire%20and%20Rutland.pdf

Gammans, N. (2008). *Conserving the red-barbed ant (*Formica rufibarbis*) in the United Kingdom*. Project Report 2008. Woking: Surrey Wildlife Trust. http://hymettus.org.uk/downloads/F%20rufibarbis%20tech%20report.pdf

Gavilanez-Slone, J. & Porter, S.D. (2014) Laboratory fire ant colonies (*Solenopsis invicta*) fail to grow with Bhatkar diet and three other artificial diets. *Insectes Sociaux* 61: 281–287. https://doi.org/10.1007/s00040-014-0353-7

Gilbertson, D.D., Kent, M. & Pyatt, F.B. (1985) *Practical Ecology for Geography and Biology: Survey, Mapping and Data Analysis*. Boston, MA: Springer. https://doi.org/10.1007/978-1-4684-1415-8_8

Golley, F.B. & Gentry, J.B. (1964) Bioenergetics of the Southern Harvester Ant, *Pogonomyrmex badius*. *Ecology* 45: 217–225. https://doi.org/10.2307/1933834

Gorb, S., Gorb, E. & Sindarovskaya, Y. (1997) Interaction between the non-myrmecochorous herb *Galium aparine* and the ant *Formica polyctena*. *Plant Ecology* 131: 215–221. https://doi.org/10.1023/A:1009789202189

Gosswald, K. & Bier, K. (1953) Untersuchungen zur Kastendetermination in der Gattung *Formica*. 2. Die Aufzucht von Geschlechtstieren bei *Formica rufa pratensis* (Retz). *Zoologischer Anzeiger* 151: 126–134

Gould, J.L. & Gould, C.G. (1988) *The Honey Bee*. New York: Scientific American Library.

Guerrieri, F.J., Nehring,V., Jørgensen, C.G., Nielsen, J., Galizia, C.G. & d'Ettorre, P. (2009) Ants recognize foes and not friends. *Proceedings of the Royal Society, London B* 276: 2461–2468. https://doi.org/10.1098/rspb.2008.1860

Guillem, R.M., Drijfhout, F. & Martin, S.J. (2014) Chemical deception among ant social parasites. *Current Zoology* 60: 62–75.

Hamer, M.T. & Cocks, L.R. (2016) *Linepithema iniquum* (Mayr) (Hymenoptera: Formicidae) found at the National Botanic Garden of Wales. *British Journal of Entomology and Natural History* 33: 71–75.

Hamer, M.T., Marquis, A.D., Guénard, B. (2021) *Strumigenys perplexa* (Smith, 1876) (Formicidae, Myrmicinae) a new exotic ant to Europe with establishment in Guernsey, Channel Islands. *Journal of Hymenoptera Research* 83: 101–124. https://doi.org/10.3897/jhr.83.66829

Hames, C.A.C. (1987) Provisional atlas of the association between *Platyarthrus hoffmannseggi* and ants in Britain and Ireland. *Isopoda* 1: 9–19.

Hamilton, W.D. (1964) The genetical evolution of social behaviour. *Journal of Theoretical Biology* 7: 1–52. https://doi.org/10.1016/0022-5193(64)90038-4

Hamilton, W.D. (1972) Altruism and related phenomena, mainly in social insects. *Annual Review of Ecology and Systematics* 3: 193–232. https://doi.org/10.1146/annurev.es.03.110172.001205

Hangartner, W. (1967) Spezifität und Inaktivierung des Spurpheromons von *Lasius fuliginosus* Latr. und Orientierung der Arbeiterinnen im Duftfeld. *Zeitschrift für Vergleichende Physiologie* 57: 103–136.

Harde, K.W. & Severa, F. (1984) *Beetles* (English edition). London: Octopus Books.

Hargreaves, B. & White, S. (2021) *The Aculeate Hymenoptera (Bees, Wasps and Ants) of Lancashire and North Merseyside*. Rishton: Lancashire and Cheshire Fauna Society.

Harvey, P.R. (1998) The modern distribution of ants in Essex with their regional rarity and threat status. *Essex Naturalist* 15: 61–111. https://www.essexfieldclub.org.uk/portal.php/p/Archive/s/114/o/0001

Hayes, A. (2018) *Ant Farms – The Ultimate Formicarium Handbook: Detailed Step-by-Step Guide to Setting Up a Thriving Ant Colony*. Scotts Valley, California: Createspace Independent Publishing Platform.

Heads, P.A. (1986) Bracken, ants and extrafloral nectaries. IV. Do wood ants (*Formica lugubris*) protect the plant against insect herbivores? *Journal of Animal Ecology* 55: 795–809. https://doi.org/10.2307/4417

Henderson, P.A. (2021) *Southwood's Ecological Methods*. 5th edition. Oxford: Oxford University Press.

Heyde, K. (1924) Die Entwicklung der psychischen Fähigkeiten bei Ameisen und ihr Verhalten bei abgeänderten biologischen Bedingungen. *Biologisches Zentralblatt* 44: 623–654.

Hickman, J.C. (1974) Pollination by ants: a low-energy system. *Science* 184: 1290–1292. https://doi.org/10.1126/science.184.4143.1290

Hodge, P.J. & Jones, R.A. (1995) *New British Beetles: Species Not in Joy's Practical Handbook*. Reading: British Entomological and Natural History Society.

Hölldobler, K. (1953) Beobachtungen über die Koloniengründung von *Lasius umbratus umbratus* Nyl. *Zeitschrift für angewandte Entomologie* 34: 598–606. https://doi.org/10.1111/j.1439-0418.1953.tb00704.x

Hölldobler, B. & Bartz, S.H. (1985) Sociobiology of reproduction in ants. *Fortschritte der Zoologie* 3: 237–257.

Hölldobler, B. & Kwapich, C.L. (2017) *Amphotis marginata* (Coleoptera: Nitidulidae) a highwayman of the ant *Lasius fuliginosus*. *PLoS ONE* 12(8): e0180847. https://doi.org/10.1371/journal.pone.0180847

Hölldobler, B. & Kwapich, C.L. (2022) *The Guests of Ants: How Myrmecophiles Interact with Their Hosts*. Harvard: Belknap Press. https://doi.org/10.4159/9780674276451

Hölldobler, B. & Wilson, E.O. (1990) *The Ants*. Berlin: Springer-Verlag. https://doi.org/10.1007/978-3-662-10306-7

Hölldobler, B. (1979) Territories of the African weaver ant (*Oecophylla longinoda*): a field study. *Zeitschrift für Tierpsychologie* 51: 201–213. https://doi.org/10.1111/j.1439-0310.1979.tb00683.x

Holman, L., Jørgensen, C.G., Nielsen, J. & d'Ettorre, P. (2010) Identification of an ant queen pheromone regulating worker sterility. *Proceedings of the Royal Society B Biological Sciences* 277: 3793–3800. https://doi.org/10.1098/rspb.2010.0984

Holt, S.J. (1955) On the foraging activity of the wood ant. *Journal of Animal Ecology* 24: 1–34. https://doi.org/10.2307/1877

Horstmann, K. & Schmidt, H. (1986) Temperature regulation in nests of the wood ant *Formica polyctena* (Hymenoptera: Formicidae). *Entomologia Generalis* 11: 229–236. https://doi.org/10.1127/entom.gen/11/1986/229

Huber, P. (1810). *Recherches sur les Moeurs des Fourmis Indigènes*. Paris: J.J. Paschoud. https://doi.org/10.5962/bhl.title.49948

Hoyt, S. (1998) *The Earth Dwellers*. Edinburgh and London: Mainstream Publishing.

Hughes, J. (2006). A review of wood ants (Hymenoptera: Formicidae) in Scotland. Scottish Natural Heritage Commissioned Report No. 178 (ROAME No. F04AC319).

Hughes, J. (1997) Review of the distribution of *Formica exsecta* Nylander in Scotland with a survey of sites with no recent records. Scottish Natural Heritage Commissioned Research Report 81.

Jaisson, P. (1991) Kinship and fellowship in ants and social wasps. In: Hepper, P.G. (ed.). *Kin Recognition* (pp. 60–93).

Cambridge: Cambridge University Press. https://doi.org/10.1017/CBO9780511525414.005

Janzen, D.J. (1967). Interaction of the bull's-horn acacia (*Acacia cornigera* L.) with an ant inhabitant (*Pseudomyrmex ferruginea* F. Smith) in eastern Mexico. *University of Kansas Science Bulletin* 47: 315–558.

Johansson, T. & Gibb, H. (2016) Interspecific competition and coexistence between wood ants. In: Stockan, J.A. & Robinson, E.J.H. *Wood Ant Ecology and Conservation* (pp. 106–122). Cambridge: Cambridge University Press. https://doi.org/10.1017/CBO9781107261402.007

Jones, N. & Cheeseborough, I. (2014) *A Provisional Atlas of Bees, Wasps and Ants of Shropshire.* Telford: Field Studies Council.

Jones, R. (2022) *Ants.* London: Bloomsbury.

Joy, N.H. (1976) *A Practical Handbook of British Beetles.* Facsimile of the 1932 edition. Faringdon: E.W. Classey.

Kadochová, S., Frouz, J.R. & Roces, F. (2017). Sun basking in red wood ants *Formica polyctena* (Hymenoptera, Formicidae): individual behaviour and temperature-dependent respiration rates. *PLoS ONE* 12: e0170570. https://doi.org/10.1371/journal.pone.0170570

Keller, R.A. & Peeters, C. (2020) Poneroid ants. In: Starr, C. (ed.) *Encyclopedia of Social Insects.* New York: Springer, Cham.

Keller, L., Cherix, D. & Ulloa-Chacon, P. (1989) Description of a new artificial diet for rearing ant colonies as *Iridomyrmex humilis, Monomorium pharaonis* and *Wasmannia auropunctata* (Hymenoptera; Formicidae). *Insectes Sociaux* 36: 348–352. https://doi.org/10.1007/BF02224886

King, T.J. (1977a) The plant ecology of ant-hills in calcareous grasslands. I. Patterns of species in relation to ant-hills in southern England. *Journal of Ecology* 65: 235–256. https://doi.org/10.2307/2259077

King, T.J. (1977b) The plant ecology of ant-hills in calcareous grasslands: II. Succession on the mounds. *Journal of Ecology* 65: 257–278. https://doi.org/10.2307/2259078

King, T.J. (1977c) The plant ecology of ant-hills in calcareous grasslands: III. Factors affecting the population sizes of selected species. *Journal of Ecology* 65: 279–315. https://doi.org/10.2307/2259079

King, T.J. (1981) Ant-hill vegetation of acidic grasslands in the Gower peninsula, South Wales. *New Phytologist* 88: 559–571. https://doi.org/10.1111/j.1469-8137.1981.tb04100.x

Kleinjan, J.E. & Mittler, T.E. (1975) A chemical influence of ants on wing development in aphids. *Entomologia Experimentalis et Applicata* 18: 384–388. https://doi.org/10.1111/j.1570-7458.1975.tb00411.x

Kloft, W.J. (1959) Versuch einer Analyse der trophobiotischen Beziehungen von Ameisen und Aphiden. *Biologisches Zentralblatt* 78: 863–870.

Kondoh, M. (1968) Bioeconomic studies on the colony of an ant species, *Formica japonica* Motschulsky. 1. Nest structure and seasonal change of the colony members. *Japanese Journal of Ecology* 18: 124–133.

Koptur, S. & Lawton, J.H. (1988) Interactions among vetches bearing extrafloral nectaries, their biotic protection agents, and herbivores. *Ecology* 69: 278–283. https://doi.org/10.2307/1943183

Kunkel, H. (1973) Die Kotabgabe der Aphiden (Aphidina, Hemiptera) unter Einfluss von Ameisen. *Bonner Zoologische Beiträge* 24: 105–121.

Lach, L., Parr, C.L. & Abbott, L.L. (eds) (2010) *Ant Ecology.* Oxford: Oxford Academic.

Lawton, J.H. & Heads, P.A. (1984) Bracken, ants and extrafloral nectaries. 1. The components of the system. *Journal of Animal Ecology* 53: 995–1014. https://doi.org/10.2307/4673

Lebas, C., Galkowski, C, Blatrix, R. & Wegnez, P. (2019). *Ants of Britain and Europe: A Photographic Guide.* London: Bloomsbury Wildlife.

Lenoir, A., Fresneau, D., Errard, C. & Hefetz, A. (1999) The individuality and colonial identity in ants: the emergence of the social representation concept. In: Deneubourg, J.L., Pasteels, J.M. & Detrain, C. (eds). *Information Processing in Social Insects* (pp. 219–237). Basel: Birkhäuser. https://doi.org/10.1007/978-3-0348-8739-7_12

Luckhurst, T. (2017) *Ant Farms. Ant Farms Guide. Ant Farms and Raising Colonies of Ants as Pets.* Zoodoo Publishing.

Macdonald, M. (2013) *Highland Ants: Distribution, Ecology and Conservation.* Inverness: Highland Biological Recording Group.

MacGregor, E.G. (1948) Odour as a basis for orientated movement in ants. *Behaviour* 1: 267–296. https://doi.org/10.1163/156853948X00137

MacKay, W.P. (1985). A comparison of the energy budgets of three species of *Pogonomyrmex* harvester ants (Hymenoptera: Formicidae). *Oecologia* 66: 484–494. https://doi.org/10.1007/BF00379338

Maschwitz, U. & Hölldobler, B. (1970) Der Kartonnestbau bei *Lasius fuliginosus* Latr. (Hym. Formicidae). *Zeitschrift für vergleichende Physiologie* 66: 176–189. https://doi.org/10.1007/BF00297777

Martin, S.J. (2016) Colony and species recognition among the *Formica* ants. In: Stockan, J.A. & Robinson, E.J.H. *Wood Ant Ecology and Conservation* (pp. 106–122). Cambridge: Cambridge University Press. https://doi.org/10.1017/CBO9781107261402.006

Martin, S.J., Drijfhout, F.P. & Hart, A.G. (2019) Phenotypic plasticity of nest-mate recognition cues in *Formica exsecta* ants. *Journal of Chemical Ecology* 45: 735–740. https://doi.org/10.1007/s10886-019-01103-2

Moffett, M.W. (1987) Ants that go with the flow: a new method of orientation by mass communication. *Naturwissenschaften* 74: 551–553. https://doi.org/10.1007/BF00367078

Monaghan, J. (2022). *Hybridisation and Genetic Structure of Woodland Specialist Ants in Fragmented Habitat*. University of York: unpublished PhD thesis.

Morgan, E.D. (2008) Chemical sorcery for sociality: exocrine secretions of ants (Hymenoptera: Formicidae). *Myrmecological News* 11: 79–90.

Mori, A. & Le Moli, F. (1998) Mating behavior and colony founding of the slave-making ant *Formica sanguinea* (Hymenoptera: Formicidae). *Journal of Insect Behavior* 11: 235–245. https://doi.org/10.1023/A:1021048024219

Munns, L., Cantarello, E. & Harrison, A. (2018) *New Forest HLS Scheme Specialist Habitat and Species Surveys: Survey and Assessment of Long-Spined Ant* Temnothorax interruptus *Status in the New Forest*. BU Global Environmental Solutions (BUG) report (BUG2774) to Forestry Commission. Higher Level Stewardship Agreement, The Verderers of the New Forest. AG00300016. Lyndhurst: Forestry Commission.

NatureScot (2020) Scottish Biodiversity List. https://www.nature.scot/sites/default/files/2022-04/Scottish%20Biodiversity%20List.xls

Niechoj, R. (2011) *Irish Ants (Hymenoptera, Formicidae): Distribution, Conservation and Functional Relationships*. University of Limerick: unpublished PhD thesis.

Nielsen, M.G. (1972) An attempt to estimate energy flow through a population of workers of *Lasius alienus* (Först) (Hymenoptera: Formicidae). *Natura Jutlandica* 16: 99–107. https://www.jstor.org/stable/4217659

Noda, C., Fernández, J., Pérez-Penichet, C. & Altshuler, E. (2006) Measuring activity in ant colonies. *Review of Scientific Instruments* 77: 126102. https://doi.org/10.1063/1.2400215

Nonacs, P. (1986) Ant reproductive strategies and sex allocation theory. *Quarterly Review of Biology* 61: 1–21. https://doi.org/10.1086/414723

Otto, D. (1958) Über die Arbeitsteilung im Staate von *Formica rufa rufo-pratensis* minor Gössw. und ihre verhaltensphysiologischen Grundlagen, ein Beitrag zur Biologie der roten Waldameise. *Wissenschaftliche Abhandlungen der Deutschen Akademie der Landwirtschaftswissenschaften zu Berlin* 30: 1–169.

Parr, C.L. & Bishop, T.R. (2022). The response of ants to climate change. *Global Change Biology* 28: 3188–3205. https://doi.org/10.1111/gcb.16140

Peakall, R.S., Handel, N. & Beattie, A.J. (1991) The evidence for, and importance of ant pollination. In: Huxley, C.R. & Cutler, D. (eds). *Ant–Plant Interactions*. Oxford: Oxford University Press.

Pearce-Higgins, J.W., Ausden, M.A., Beale, C.M., Oliver, T.H. & Crick, H.Q.P. (eds) (2015). Research on the assessment of risks & opportunities for species in England as a result of climate change. Natural England Commissioned Reports, Number 175.

Peeters, C. & Molet, M. (2010) Colonial reproduction and life histories. In: Lach, L., Parr, C.L. & Abbott, L.L. (eds). *Ant Ecology*. Oxford: Oxford University Press. https://doi.org/10.1093/acprof:oso/9780199544639.003.0009

Perkins, R.C.L. (1913) Introduction. In: Sharp, D. (ed.). *Fauna Hawaiiensis*. Cambridge: Cambridge University Press.

Pontin, J. (2005) *Ants of Surrey*. Woking: Surrey Wildlife Trust.

Plowman, J. (1995) *Biodiversity: The UK Steering Group Report. Volume 1: Meeting the Rio Challenge*. Peterborough: Joint Nature Conservation Committee.

Puterbaugh, M. (1996) The roles of ants as flower visitors: experimental analysis in three alpine plant species. *Oikos* 83: 36–46. https://doi.org/10.2307/3546544

Radchenko, A.G. (2016). *Ants (Hymenoptera, Formicidae) of Ukraine*. Kyiv: National Academy of Sciences of Ukraine, Schmalhausen Institute of Zoology.

Rashbrook, V.K., Compton, S.G. & Lawton, J.H. (1992) Ant–herbivore interactions:

reasons for the absence of benefits to a fern with foliar nectaries. *Ecology* 73: 2167–2174. https://doi.org/10.2307/1941464.

Rees, S.D. (2006). *Conservation Genetics and Ecology of the Endangered Black Bog Ant*, Formica picea. Ann Arbor: ProQuest. https://orca.cardiff.ac.uk/id/eprint/56197/1/U584972.pdf

Robinson, N.A. (1998) Two new records of the myrmecophile spider *Thyreosthenius biovatus* Cambridge in nests of *Formica rufa* L. *British Journal of Entomology and Natural History* 11: 72.

Robinson, N.A. (2005) A list of the bees, wasps and ants of Cumbria, Records to the year 2005. *BWARS Newsletter*, Autumn 2005: 25–30.

Robinson, E.J.H. (2011) *A Survey of Changes in Wood Ant Distribution in Northern England and the Midlands. Hymettus Survey Report: Changes in Wood Ant Distribution*. Midhurst: Hymettus.

Robinson, E.J.H., Stockan, J.A. & Iason, G.R. (2016) Wood ants and their interactions with other organisms. In: Stockan, J.A. & Robinson, E.J.H. (eds). *Wood Ant Ecology and Conservation*. Cambridge: Cambridge University Press. https://doi.org/10.1017/CBO9781107261402.009

Rosengren, R. (1971) Route fidelity, visual memory and recruitment behaviour in foraging wood ants of genus *Formica* (Hymenoptera, Formicidae). *Acta Zoologica Fennica* 133: 1–106.

Rosengren, R. (1987) Polyethic structure of the foraging/guarding system of red wood ants (*Formica* s. str.). In: Eder, J. & Rembold, H. (eds) *Chemistry and Biology of Social Insects. Proceedings of the Tenth International Congress of the International Union for the Study of Social Insects* (pp. 118–119). Munich: J. Peperny.

Rotheray, G. (1989) *Aphid Predators*. Naturalists' Handbooks 11. Slough: The Richmond Publishing Co. Ltd.

Rotheray, G. (1994) *Insect Life on Plants*. London: Chapman and Hall.

Salem, M. & Hole, F.D. (1968) Ant (*Formica cinerea*) pedoturbation in a forest soil. *Soil Science Society of America Proceedings* 32: 563–567. https://doi.org/10.2136/sssaj1968.03615995003200040039x

Savolainen, R. & Vepsäläinen, K. (1988). A competition hierarchy among boreal ants: impact on resource partitioning and community structure. *Oikos* 51: 135–155. https://doi.org/10.2307/3565636

Scherba, G. (1959) Moisture regulation in mound nests of the ant, *Formica ulkei* Emery. *American Midland Naturalist* 61: 499–508. https://doi.org/10.2307/2422517

Schultheiss, P., Nooten, S.S., Wang, R., Wong, M.K.L., Brassard, F. & Guénard, B. (2022). The abundance, biomass, and distribution of ants on Earth. *PNAS* 119: 1–9. https://doi.org/10.1073/pnas.2201550119

Schwenke, W. (1957) Über die räuberische Tätigkeit von *Formica rufa* L. und *Formica nigricans* Emery außerhalb einer Insekten-Massenvermehrung. *Beiträge zur Entomologie* 7: 226–246.

Seeley, T.D. (1985) *Honeybee Ecology*. Princeton, NJ: Princeton University Press.

Seifert, B. (2000) A taxonomic revision of the ant subgenus *Coptoformica* Mueller, 1923 (Hymenoptera, Formicidae). *Zoosystema* 22: 517–568.

Seifert, B. (2012) Clarifying naming and identification of the outdoor species of the ant genus *Tapinoma* Förster, 1850 (Hymenoptera: Formicidae) in Europe north of the Mediterranean region with description of a new species. *Myrmecological News* 16: 139–147.

Seifert, B. (2018) *The Ants of Central and Northern Europe*. Boxberg/Oberlausitz, Tauer: Lutra Verlags- und Vertriebsgesellschaft.

Seifert, B., Kulmuni, J., Pamilo, P. (2010) Independent hybrid populations of *Formica polyctena X rufa* wood ants (Hymenoptera: Formicidae) abound under conditions of forest fragmentation. *Evolutionary Ecology* 24: 1219–1237. https://doi.org/10.1007/s10682-010-9371-8

Shaw, M.R. & Huddleston, T. (1991) *Classification and Biology of Braconid Wasps (Hymenoptera: Braconidae)*. Handbooks for the Identification of British Insects. Vol. 7 Pt 11. London: Royal Entomological Society.

Skaife, S.H. (1961) *The Study of Ants*. London: Longmans Green.

Skinner, G.J. (1976) *Some Aspects of the Ecology of the Wood Ant*, Formica rufa L. *(Hymenoptera: Formicidae), in Limestone Woodlands of North-West England*. University of Lancaster: unpublished PhD thesis.

Skinner, G.J. (1980a) Territory, trail structure and activity patterns in the wood-ant *Formica rufa* (Hymenoptera: Formicidae) in limestone woodland in north-west England. *Journal of Animal Ecology* 49: 381–394. https://doi.org/10.2307/4253

Skinner, G.J. (1980b) The feeding habits of the wood-ant *Formica rufa* (Hymenoptera,

Formicidae) in limestone woodland in north-west England. *Journal of Animal Ecology* 49: 417–433. https://doi.org/10.2307/4255

Skinner, G.J. (1987) *Ants of the British Isles*. Princes Risborough: Shire Publications.

Skinner, G.J. & Allen, G.W. (1996). *Ants*. Naturalists' Handbooks 24. Slough: The Richmond Publishing Co.

Skinner, G.J. & Whittaker, J.B. (1981). An experimental investigation of inter-relationships between the wood-ant (*Formica rufa*) and some tree-canopy herbivores. *Journal of Animal Ecology* 50: 313–326. https://doi.org/10.2307/4047

Shirt, D.B. (ed.) (1987) *British Red Data Books: 2. Insects*. Peterborough: Joint Nature Conservation Committee.

Sorvari, J. (2017). Threats, conservation and management. In: Stockan, J.A. & Robinson, E.J.H. (eds). *Wood Ant Ecology and Conservation*. Cambridge: Cambridge University Press. https://doi.org/10.1017/CBO9781107261402.013

Stadler, B. & Dixon, A.F.G. (2008) Ecology and evolution of aphid–ant interactions. *Annual Review of Ecology, Evolution and Systematics* 36: 345–372. https://doi.org/10.1146/annurev.ecolsys.36.091704.175531

Stockan, J.A, & Robinson, E.J.H. (eds) (2016) *Wood Ant Ecology and Conservation (Ecology, Biodiversity and Conservation)*. Cambridge: Cambridge University Press. https://doi.org/10.1017/CBO9781107261402

Styrsky, J. & Eubanks, M. (2007) Ecological consequences of interactions between ants and honeydew-producing insects. *Proceedings of the Royal Society B Biological Sciences* 274: 151–164. https://doi.org/10.1098/rspb.2006.3701

Subedi, I.P. (2016) Ants – ecosystem engineers. *KIST Newsletter* 1: 9–10.

Sudd, J. & Franks, N. (1987) *The Behavioural Ecology of Ants*. London: Chapman & Hall. https://doi.org/10.1007/978-94-009-3123-7

Sudd, J.H. (1967) *An Introduction to the Behaviour of Ants*. London: Edward Arnold.

Thomas, J. & Lewington, R. (1991) *The Butterflies of Britain and Ireland*. London: Dorling Kindersley.

Thomas, J.A., Simcox, D.J. & Clarke R.T. (2009) Successful conservation of a threatened *Maculinea* butterfly. *Science* 325: 80–83. https://doi.org/10.1126/science.1175726

Tschinkel, T. (2021) *Ant Architecture: The Wonder, Beauty, and Science of Underground Nests*. Princeton: Princeton University Press. https://doi.org/10.1515/9780691218496

Trivers, R.L. & Hare, H. (1976) Haplodiploidy and the evolution of the social insect. *Science* 191: 249–263. https://doi.org/10.1126/science.1108197

Vulliamy, B. (2020) *Narrow-Headed Ant Formica exsecta Captive Queen Rearing, Mating and Release 2020. Back from the Brink. Buglife Report*. Peterborough: Buglife.

Wagner, H.C., Arthofer, W., Seifert, B., Muster, C., Steiner, F.M. & Schlick-Steiner, B.C. (2017) Light at the end of the tunnel: Integrative taxonomy delimits cryptic species in the *Tetramorium caespitum* complex (Hymenoptera: Formicidae). *Myrmecological News* 25: 95–129. https://doi.org/10.25849/myrmecol.news_025:095

Waloff, N. & Blackith, R.E. (1962) The growth and distribution of the mounds of *Lasius flavus* (Fabricius) (Hymenoptera: Formicidae) in Silwood Park, Berkshire. *Journal of Animal Ecology* 31: 421–437. https://doi.org/10.2307/2044

Walters, J. (2020) *Narrow-Headed Ant*, Formica exsecta, *Back from the Brink. Buglife Report*. Peterborough: Buglife.

Wasmann, E. (1894) *Kritisches Verzeichniss der mymecophilen und termitophylen Arthropoden mit Angabe der Lebensweise und mit Beschreibung neuer Arten*. Berlin: Verlag Felix L. Dames

Way, M.J. (1954) Studies of the association of the ant *Oecophylla longinoda* (Latr.) (Formicidae) with the scale insect *Saissetia zanzibarensis* Williams (Coccidae). *Bulletin of Entomological Research* 45: 113–134. https://doi.org/10.1017/S0007485300026833

Weir, J.S. (1958a) Polyethism in workers of the ant *Myrmica* I. *Insectes Sociaux* 5: 97–128. https://doi.org/10.1007/BF02222431

Weir, J.S. (1958b) Polyethism in workers of the ant *Myrmica* II. *Insectes Sociaux* 5: 315–339. https://doi.org/10.1007/BF02223941

Wells, D. (2018). Ants of Norfolk: cumulative recorder report 2008–2018. *Transactions of the Norfolk and Norwich Naturalists' Society* 51: 1–3.

Wheater, C.P. & Cook, P.A. (2003) *Studying Invertebrates*. Naturalists' Handbooks 28. Exeter: Pelagic Publishing.

Wheeler, W.M. (1910) *Ants: their Structure, Development and Behavior*. New York: Columbia University Press. https://doi.org/10.5962/bhl.title.1937

White, W.F. (1895) *Ants and their Ways*. London: Religious Tract Society.

Whittaker, J.B. & Warrington, S. (1985) An experimental field study of different levels of insect herbivory induced by *Formica rufa* predation on Sycamore (*Acer pseudoplatanus*) III. Effects on tree growth. *Journal of Applied Ecology* 22: 797–811. https://doi.org/10.2307/2403230

Wiken, E.B., Broersma, K., Lavkulich, L.M. & Farstad, I. (1976) Biosynthetic alteration in a British Columbia soil by ants (*Formica fusca* Linné). *Soil Science Society of America Journal* 40: 422–426. https://doi.org/10.2136/sssaj1976.03615995004000030032x

Williams, T. & Franks, N. (1985) Population size and growth rate, sex ratio and behaviour in the ant isopod, *Platyarthrus hoffmannseggi*. *Journal of Zoology* 215: 703–717. https://doi.org/10.1111/j.1469-7998.1988.tb02405.x

Williams, T. & Franks, N. (1988) Research on *Platyarthrus hoffmannseggi* at Bath University. *Newsletter of the British Isopod Study Group* 19: 6–8.

Wilson, E.O. & Farish, D.J. (1973) Predatory behaviour in the ant-like wasp *Methocha stygia* (Say) (Hymenoptera: Tiphiidae).

Animal Behaviour 21: 292–295. https://doi.org/10.1016/S0003-3472(73)80069-7

Wiswell, H., Attewell, P., Carroll, S. & Stockan, J. (2022) *Wood Ant Translocation: Good Practice Guidance*. Grantown-on-Spey: Cairngorms National Park Authority.

Wragge Morley, D. (1953) *Ants*. London: Collins.

Wong, M.K.L., Economo, E.P. & Guénard, B. (2023). The global spread and invasion capacities of alien ants. *Current Biology* 33: 566–571. https://doi.org/10.1016/j.cub.2022.12.020

Yarrow, I.H.H. (1954) The British ants allied to *Formica fusca* L. (Hym., Formicidae). *Transactions of the Society for British Entomology* 11: 229–244.

Yarrow, I.H.H. (1955) The British ants allied to *Formica rufa* L. (Hym., Formicidae). *Transactions of the Society for British Entomology* 12: 1–48.

Zahn, M. (1958) Temperatursinn, Wärmehaushalt und Bauweise der roten Waldameisen (*Formica rufa* L.) *Zoologische Beiträge*, n.s. 3: 127–194.

Index

References to figures and photographs appear in *italic* type; those in **bold** type refer to tables.